LAB MANUAL

TO ACCOMPANY

PRACTICAL COOLING TECHNOLOGY

WILLIAM M. JOHNSON
Central Piedmont Community College
Charlotte, North Carolina

Delmar Publishers
An International Thomson Publishing Company I(T)P®

Albany • Bonn • Boston • Cincinnati • Detroit • London • Madrid
Melbourne • Mexico City • New York • Pacific Grove • Paris • San Francisco
Singapore • Tokyo • Toronto • Washington

NOTICE TO THE READER

Cover Photos courtesy of: Robinair Division, SPX Corporation, Beckman Industrial Corporation, and Bill Johnson
Cover Design by: Charles Cummings Advertising/Art, Inc.

DELMAR STAFF
Publisher: Robert D. Lynch
Senior Administrative Editor: Vernon Anthony
Developmental Editor: Denise Denisoff
Production Coordinator: Karen Smith
Art/Design Coordinator: Cheri Plasse

Printed in the United States of America

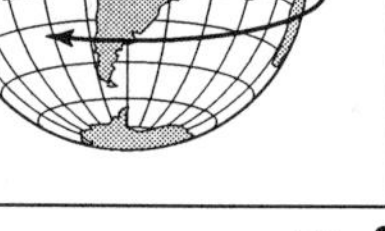

For more information, contact:

Delmar Publishers
3 Columbia Circle, Box 15015
Albany, New York 12212-5015

International Thomson Publishing Europe
Berkshire House 168-173
High Holborn
London, WC1V7AA
England

Thomas Nelson Australia
102 Dodds Street
South Melbourne, 3205
Victoria, Australia

Nelson Canada
1120 Birchmont Road
Scarborough, Ontario
Canada M1K 5G4

International Thomson Editores
Campos Eliseos 385, Piso 7
Col Polanco
11560 Mexico D F Mexico

International Thomson Publishing GmbH
Königswinterer Strasse 418
53227 Bonn
Germany

International Thomson Publishing Asia
221 Henderson Road
#05-10 Henderson Building
Singapore 0315

International Thomson Publishing Japan
Hirakawacho Kyowa Building, 3F
2-2-1 Hirakawacho
Chiyoda-ku,Tokyo 102
Japan

2 3 4 5 6 7 8 9 10 XXX 02 01 00 99 98 97

Library of Congress Catalog Card Number: 96-41093
ISBN: 0-8273-7603-0

Contents

PREFACE

The lab exercises in this manual are intended to enable the student to perform work exercises that are similar to actual field experiences. Some of the exercises are to be written only. The purpose of having the student perform calculations and answer questions is to stimulate interest while reinforcing the subject matter. Actual service exercises are included, when possible, so the student actually performs the exercises on equipment.

An effort was made to make the exercises as generic as possible. Most school systems should have the equipment used in these labs. The class may use equipment that the school uses to heat the building for some projects. No exercise in this text should cause a piece of equipment to be put out of service for any longer than it takes to complete the exercise.

A recommended tool list is included for the purpose of providing the lab instructor with the tools necessary to complete these labs. This list can also be used to help in choosing the tools to be used for the course.

Particular emphasis should be placed on the "Safety Precautions" that appear throughout the text and lab manual. Students should be required to adhere to safe lab and shop practices for their own protection and to form good safety habits.

Answers to the questions in the text and lab manual can be found in the Instructor's Guide, which accompanies *Practical Cooling Technology.* You may also refer to the Instructor's Guide for suggestions regarding organization of the material and general teaching suggestions.

TOOL LIST

This tool list includes the tools required to complete the lab exercises in this book as well as some basic supplies. Where brand names are referred to in this book, it is not meant to be an endorsement of any particular manufacturer. These tools have been used successfully for these exercises in our program.

Goggles, gloves, and calculator
Portable electric drill with drill bits
Tube cutter with reamer
Flaring tools set
Tubing brushes, 1/4", 3/8", 1/2"
Vise to hold tubing
Service valve wrench
Sling psychrometer
Oil furnace nozzle wrench
Two 8" adjustable wrenches
Two 10" and two 14" pipe wrenches
Allen wrench set
Open or box end wrenches, 3/8", 1/2", 9/16", 5/8"
Small socket wrench set
Needlenose, slip joint, and side cutting pliers
Phillips and straight blade screwdrivers
Nut drivers, 1/4" and 5/16"
Soft face hammer
Tape measure
Inspection mirror with telescoping handle
Flashlight
Appropriate manufacturer's charging charts for systems being used
Appropriate wiring diagrams for systems being used
Comfort and psychrometric charts
Wire size chart; pressure-temperature chart
Liquid line drier
2-way liquid line drier
Pressure taps
1/4" tubing tees
1/4" Schrader valve fittings
Flare nuts, 1/4", 3/8", 1/2"
Reducing flare unions, 1/4" × 3/8", 3/8" × 1/2"
Flare plugs, 1/2"
Flare unions, 1/4" x 1/4"
Plugs for gage manifold lines
Refrigerant recovery machine
DOT-approved refrigerant recovery cylinders
Cylinders of appropriate refrigerants for systems used
Soft copper ACR tubing, 1/4", 3/8", 1/2" OD
Cylinder of nitrogen with regulator
Resistance heater
Fan-limit control
Remote bulb thermostat
Gas furnace thermocouple
Room thermostat
Adjustable low-pressure control
Relay, contactor, motor starter
Thermocouple adapter
Adapters to adapt gages to oil burner pump
Several resistors of known value
20,000-ohm, 5-watt resistor
Start and run capacitor
Shallow pan
100-W trouble lights; 150-W flood lights
Light lubricating and refrigerant oil
Low temperature solder with flux, 50/50, 95/5, or low temperature silver
High temperature brazing filler metal with flux
Sandcloth (approved for hermetic compressor)
Emery cloth (300 grit)
Low-voltage wire
Electrical tape
Insulation material for insulating thermometers
Cardboard for blocking a condenser
Colored pencils
Compressed air
Thread seal compatible with natural gas
Rags
Chlorine solution
Matches

LAB 1 Flaring Tubing

Name ______________________________ Date ______________ Grade ________

OBJECTIVES: Upon completion of this exercise, you will be able to cut copper tubing, make a flare connection, and perform a simple leak check, to ensure that the tubing connections are leak-free and that they may be used for refrigerant.

INTRODUCTION: You will cut tubing of different sizes, make a flare on the end, assemble it with fittings into a single piece, pressurize it, and check it for leaks.

TEXT REFERENCES: Unit 4.

TOOLS AND MATERIALS: A tube cutter with reamer, a flaring tool set, refrigerant oil, and leak detector (halide or electronic), goggles, a gage manifold, a cylinder of nitrogen, soap bubbles, and the following materials:

2 each flare nuts, 1/4", 3/8" and 1/2"
1 each reducing flare unions, 1/4" × 3/8" and 3/8" × 1/2"
1 each 1/2" flare plug
1 each 1/4" × 1/4" flare union
Soft copper tubing 1/4", 3/8" and 1/2" OD
See Figure 1 for a photo showing these materials.

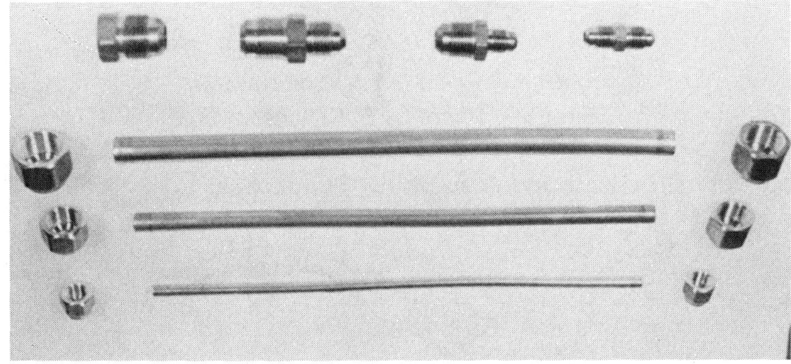

Figure 1 *Photo by Bill Johnson*

SAFETY PRECAUTIONS: Care must be used while cutting and flaring tubing. Your instructor should teach you to use the tube cutter, reamer, flaring tools, leak detector and gage manifold before you begin this exercise. Wear goggles while transferring refrigerant and pressure testing the assembly.

PROCEDURES

1. Cut a ten-inch length of each of the tubing sizes, 1/4", 3/8" and 1/2".

2. Ream (carefully) the end of each piece of tubing to prepare it for flaring, Figure 2.

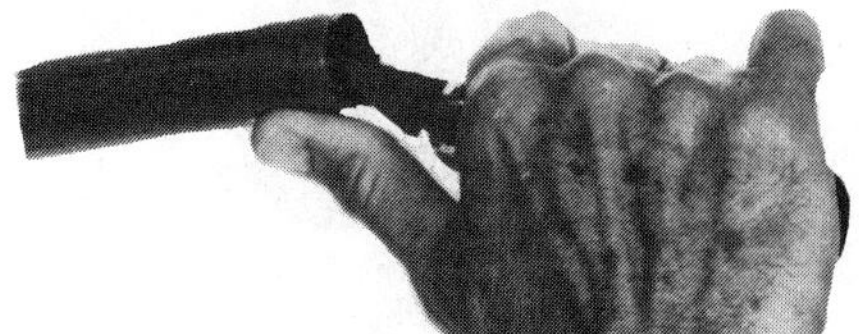

Figure 2 *Photo by Bill Johnson*

3. Slide the respective flare nuts over each piece of tubing before the flare is made, Figure 3. Be sure to apply some refrigerant oil to the flare cone before the flare is made. Make the flare.

4. Assemble the connections as in Figure 4 and tighten all connections.

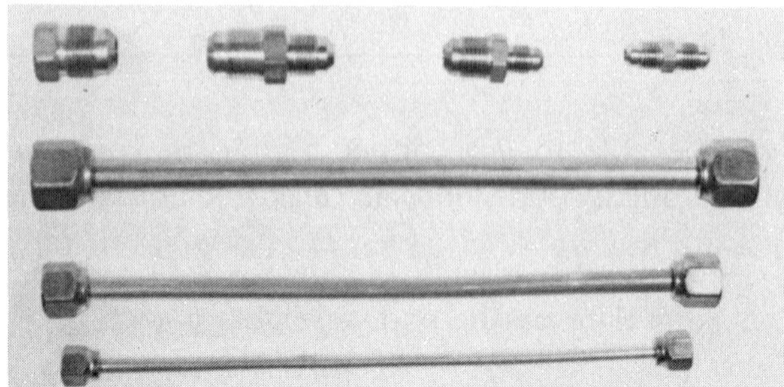

Figure 3 *Photo by Bill Johnson*

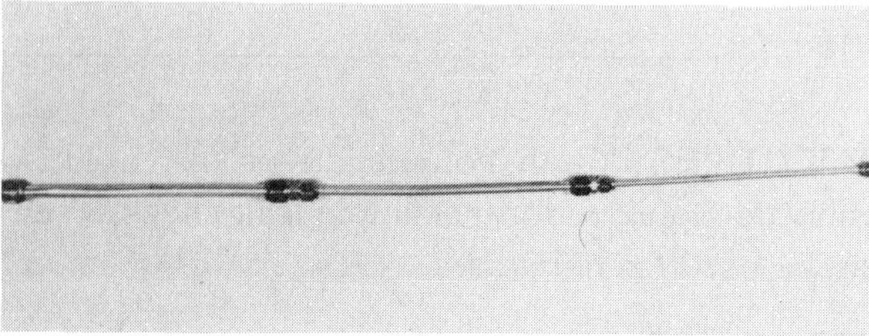

Figure 4 *Photo by Bill Johnson*

5. While wearing goggles, apply pressure to the assembly with the cylinder of nitrogen and the gage manifold, Figure 5.

Figure 5 *Photo by Bill Johnson*

6. Leak test the assembly very slowly around each fitting, Figure 6.

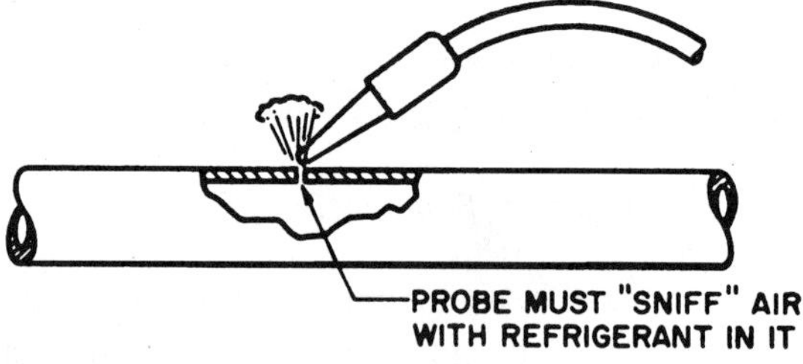

Figure 6

7. Examine any connection that leaks and repair either by replacing the brass fitting or flaring the tubing again. A scratch on the brass fitting may be a possible leak. A ridge around the flare is a sign that the tubing was not reamed correctly and will probably leak, Figure 7.

Figure 7 *Photo by Bill Johnson*

8. After all fittings have been proven leak-free, take the assembly apart except for the end plug and the 1/4" × 1/4" flare union at the ends. Cut all flare nuts off except the two on the end. They will be used for pressure testing the next exercise. Save the pieces of tubing for the next exercise, which is soldering.

MAINTENANCE OF WORK STATION AND TOOLS: Wipe any oil off the tools and return all tools and supplies to their places. Store the leftover tubing lengths for the next exercise.

SUMMARY STATEMENT: Describe, step by step, the flaring process.

QUESTIONS

1. Why should you apply refrigerant oil to the flare cone before making a flare?

2. What happens if the tubing cutter is tightened down too fast while cutting?

3. What is the purpose of reaming the tube before flaring?

4. Why is thread sealing compound not used on flare connections?

5. What actually seals the connection in a flare connection?

6. Why is a flare connection sometimes preferred to a solder connection?

7. What is the angle on a flare connection?

8. Give two reasons why a leak may occur in a flare connection.

9. Name two methods of leak testing a flare connection.

10. When a flare connection leaks due to a bad flare, what is the recommended repair?

LAB 2 Soldering and Brazing

Name ______________________________ Date ______________ Grade ________

OBJECTIVES: Upon completion of this exercise, you will be able to make solder connections with copper tubing using both low-temperature and high-temperature solder.

INTRODUCTION: You will first join together the tubing used in the previous exercise using low-temperature solder, then cut the connections apart and join them using high-temperature solder.

TEXT REFERENCES: Unit 4.

TOOLS AND MATERIALS: An air acetylene unit with a medium tip, a lighter, clear safety goggles, light gloves, low-temperature solder and flux (50/50, 95/5 or low-temperature silver), high-temperature brazing filler metal and the proper flux, sand cloth (approved for hermetic compressors) or wire brushes (1/4", 3/8" and 1/2"), a vise to hold the tubing, two each reducing couplings (sweat type) 1/4" × 3/8", 3/8" × 1/2", a leak detector, a manifold gage, and a cylinder of refrigerant.

SAFETY PRECAUTIONS: You will be working with acetylene gas and heat. READ THE TEXT before starting. Your instructor must also provide you with instruction in soldering and brazing before you begin this exercise. Wear goggles and light gloves while soldering or brazing, and while transferring refrigerant.

PROCEDURES

1. Clean the ends of the tubing with the sand cloth.

2. Clean the inside of the fittings using sand cloth or brushes.

3. Fit the assembly together, Figure 1.

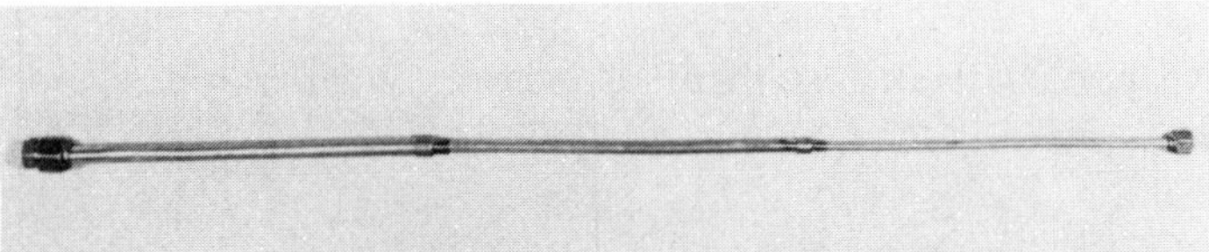

Figure 1 *Photo by Bill Johnson*

4. Place the assembly in a vertical position using the vise. This will give you practice in soldering both up and down in the vertical position, Figure 2.

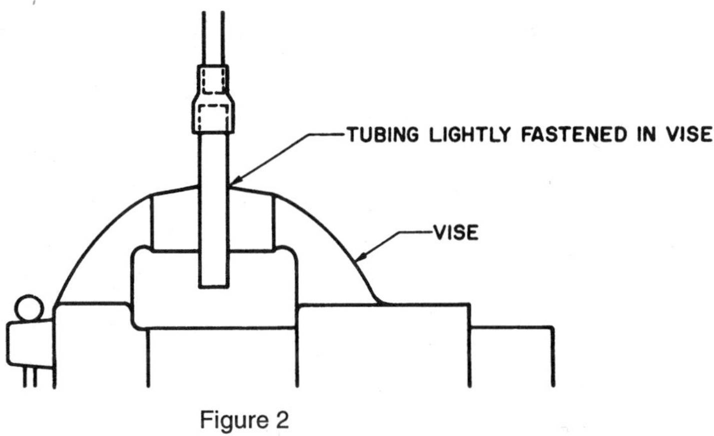

Figure 2

5. Put on your goggles. Start with the top joint at the top connection, then solder the joint underneath, Figure 3. After completing this, solder the top joint of the second connection, then solder underneath. Be sure not to overheat the connections. IF THE TUBING GETS RED OR DISCOLORS BADLY, THE CONNECTION IS TOO HOT.

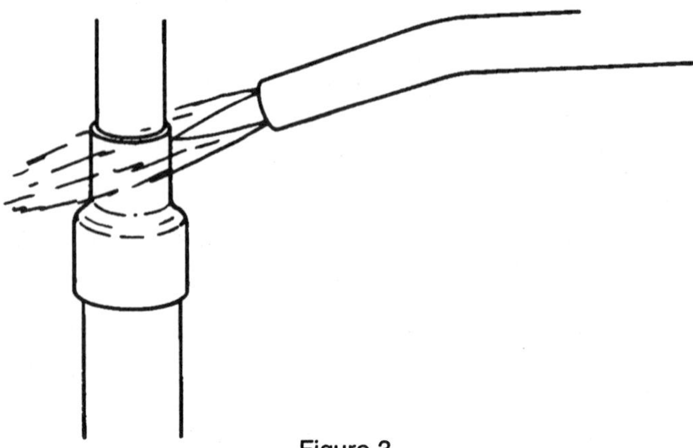

Figure 3

6. Let the assembly cool before you disturb it or the connections may come loose.

7. After the assembly has cooled, perform a leak test as in Lab 1. You may also want to submerge the pressurized assembly in water for a test.

8. Cut the reducing connectors out of the assembly and reclean the tubing ends.

9. Clean the second set of reducing couplings.

10. Assemble the pipes and connectors again and hold them upright in the vise.

11. Put on your goggles, solder the top connections, then the bottom connections as in the previous exercise, only use high-temperature brazing material. CAUTION: THE PIPE AND FITTINGS WILL HAVE TO BE CHERRY RED TO MELT THE FILLER METAL, SO BE CAREFUL.

12. After the fittings have cooled, leak test them as before. Wear goggles and gloves while transferring refrigerant.

13. Upon completion of the leak test, cut off the flare nuts and scrap the tubing.

MAINTENANCE OF WORK STATION AND TOOLS: CLEAN ANY FLUX OFF THE TOOLS OR FITTINGS AND THE WORK BENCH. Turn off the torch and bleed the acetylene from the torch hose. Place all extra fittings and tools in their proper places.

SUMMARY STATEMENTS: Describe the advantages and disadvantages of low-temperature and high-temperature solder.

QUESTIONS

1. What is the approximate melting temperature of 50/50 solder?

2. What is the approximate melting temperature of 45% silver solder?

3. What is the approximate melting temperature of 15% silver solder?

4. Should the flame or the tube melt the solder in the soldering process?

5. What is the metal content of 50/50 solder?

6. What is the metal content of 95/5 solder?

7. Which solder is stronger, 95/5 or 15% silver solder? Why?

8. Which solder would be the best choice for the discharge line, 95/5 or silver solder? Why?

9. What pulls the solder into the connection when it is heated? Explain this process.

10. Why is a special sand cloth used on hermetic compressor systems?

Lab 3 Standing Pressure Test

Name ______________________________ Date ______________ Grade ________

OBJECTIVES: Upon completion of this exercise, you will be able to perform a standing pressure test on a vessel using dry nitrogen.

INTRODUCTION: Before a gage manifold may be used with confidence in a high vacuum situation, it must be leak-free. You will use dry nitrogen to test your gage manifold to assure that it is leak-free. The standing pressure test is the same for any vessel, including a refrigeration system.

TEXT REFERENCES: Unit 5.

TOOLS AND MATERIALS: A gage manifold, plugs for the gage manifold lines, a cylinder of dry nitrogen, and goggles.

SAFETY PRECAUTIONS: Goggles should be worn at all times while working with pressurized gases. NEVER USE DRY NITROGEN WITHOUT A PRESSURE REGULATOR. THE PRESSURE IN THE TANK CAN BE IN EXCESS OF 2000 PSI. Your instructor must give you instruction in the use of the gage manifold and a dry nitrogen setup before you start this exercise.

PROCEDURES

1. Check the gages for calibration. With the gage opened to the atmosphere, it should read 0 psig. If it does not, use the calibration screw under the gage glass and calibrate it to 0 psig.

2. Check the gaskets in both ends of the gage hoses. These are often tightened too much and the gaskets ruined. If the gasket is defaced, replace it. Figure 1 shows a good and defective gasket.

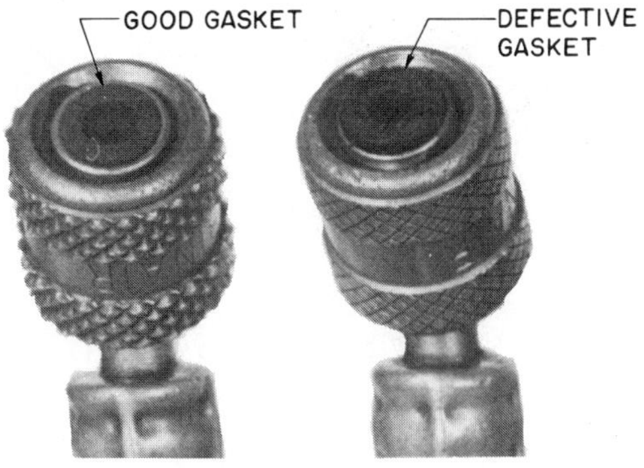

Figure 1 *Photo by Bill Johnson*

3. Fasten the center gage line to the dry nitrogen regulator and open both gage manifold handles. This opens the high- and low-side gages to the center port, Figure 2.

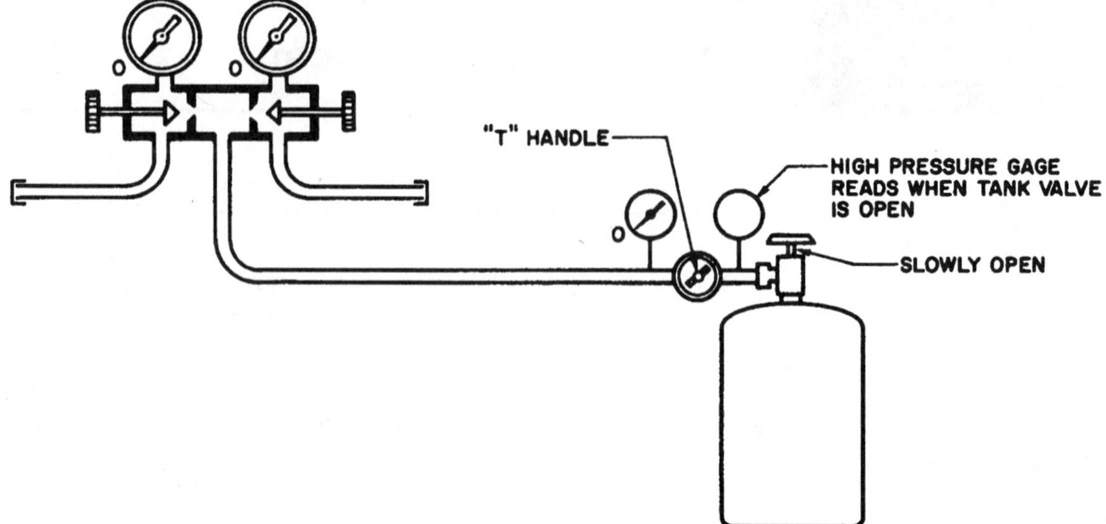

Figure 2

4. Place plugs in the high- and low-side gage lines. CAUTION: TURN THE NITROGEN REGULATOR 'T' HANDLE ALL OF THE WAY OUT, Figure 2. Now slowly open the tank valve and allow tank pressure into the regulator. DO NOT STAND IN FRONT OF THE REGULATOR 'T' HANDLE.

5. Slowly turn the 'T' handle inward and allow pressure to fill the gage manifold. Allow the pressure in the manifold to build up to 150 psig, then shut off the tank valve and turn the 'T' handle outward. The regulator is now acting as a plug for the center gage line. See Figure 3.

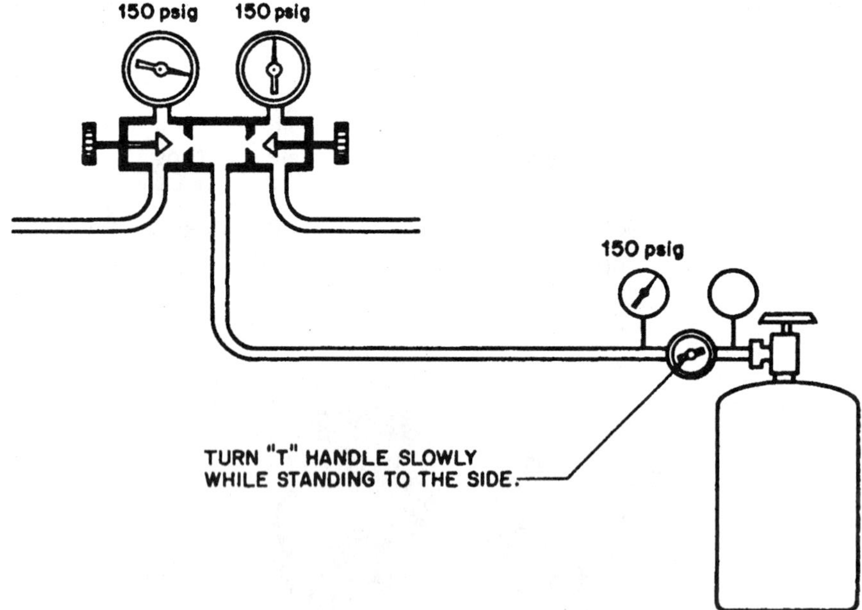

Figure 3

6. Allow the pressurized gage manifold to set, undisturbed for 15 minutes. THE PRESSURE SHOULD NOT DROP.

7. If the pressure drops, pressurize the manifold again and valve off the manifold by closing the gage manifold valves (turn toward center). A drop in pressure in either gage will help to locate the leak.

NOTE: If you cannot make the manifold leak-free by gasket replacement, you may want to submerge the gage lines in water to determine where the pressure is escaping. Gage lines often leak around the knurled ends or become porous and need to be replaced.

8. If you cannot stop the leak in a gage manifold, replace the gage lines with 1/4" copper lines used in evacuation of a system.

MAINTENANCE OF WORK STATION AND TOOLS: Place all tools in their proper places. Your instructor may suggest that you leave the gage manifold under pressure for a period of time by placing plugs in the lines. If you do this, you will know whether the gage manifold is leak-free the next time you use it.

SUMMARY STATEMENT: Describe the reason that dry nitrogen is used in a pressure test instead of a refrigerant.

Describe why a leak test while the system is in a positive pressure is more effective than using a vacuum for leak checking.

QUESTIONS

1. What can cause a gage line to leak?

2. What is the gage manifold made of?

3. What fastens the gage line to the end fittings on the end of a gage line?

4. What seals the gage line fitting to the gage manifold?

5. What can be done for repair if a gage line leaks at the fitting?

6. What would happen if a leaking gage manifold were used while trying to evacuate a system to a deep vacuum?

7. Why is refrigerant not used for a standing pressure test?

8. How much pressure does the atmosphere exert on a gage line under a deep vacuum?

9. How is a gage line tightened on a fitting?

10. What may be done for gage lines if the flexible ones cannot be made leak-free?

LAB 4 Performing a Deep Vacuum

Name ______________________________ Date ______________ Grade ______

OBJECTIVES: Upon completion of this exercise, you will be able to perform a deep vacuum test using a high-quality vacuum pump and an electronic vacuum gage.

INTRODUCTION: You will perform a high-vacuum (200 microns) evacuation on a typical refrigeration system that has service valves. A high-vacuum evacuation is the same whether the system is small or large. The only difference is the time it takes to evacuate the system. Larger vacuum pumps are typically used on large systems to save time. It is always best to give a system a standing pressure test using nitrogen before a deep vacuum is pulled to assure that the system is leak-free. An overnight pressure test of 150 psig with no pressure drop will assure that the vacuum test will go as planned. Performing a vacuum test on just the gage manifold and gage lines will also help assure that the test will be performed satisfactorily.

TEXT REFERENCES: Unit 5.

TOOLS AND MATERIALS: A leak-free gage manifold, a cylinder of refrigerant, a high-quality vacuum pump (two-stage is preferred), an electronic vacuum gage with some arrangement for valving it off from the system, a refrigerant recovery system, goggles, and gloves.

SAFETY PRECAUTIONS: Goggles and gloves are safety requirements while transferring refrigerant. A valve arrangement such as in Figure 1 may be used to protect the electronic vacuum gage element from pressure. The electronic vacuum gage sensing element must be protected from any refrigerant oil that may enter it. This may be accomplished by mounting it with the fitting end down, Figure 2. Your instructor must give you instruction in using a gage manifold, refrigerant recovery system, vacuum pump, and an electronic vacuum gage before starting this exercise.

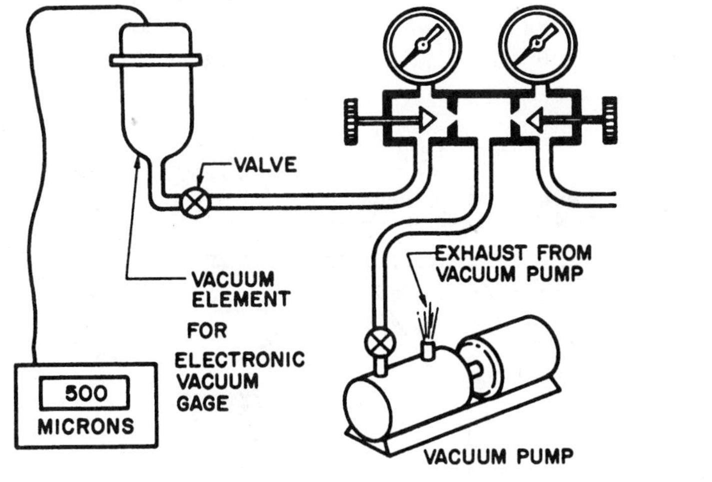

Figure 1

ELECTRICAL CONNECTION TO INSTRUMENT
GAGE CONNECTION
VALVE TO PROTECT ELEMENT

Figure 2

PROCEDURES

1. Fasten the gage manifold lines to the refrigeration system to be evacuated. Fasten the lines to the service valves at the high and the low side of the system if possible. DO NOT overtighten the gage line connections or you will damage the gaskets.

2. Position the service valve stems in the mid-position and tighten the packing glands on the valve stems, Figure 3. Be sure to replace the protective caps before starting the evacuation. Many service technicians have been deceived because of leaks at the service valves. If the system has Schrader valves, you may want to remove the valve stems during evacuation and replace them after completing the test. This will speed the evacuation procedure.

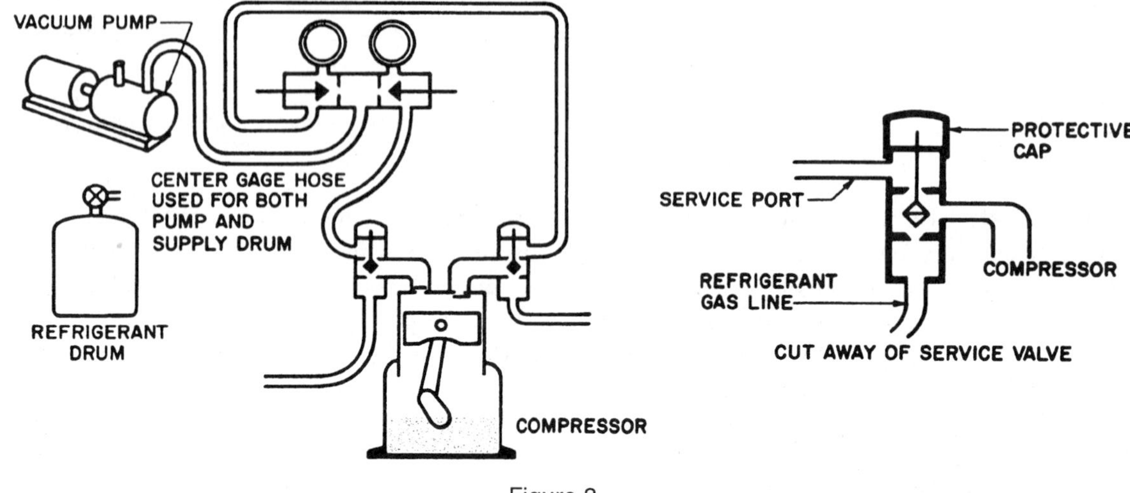

Figure 3

3. If the system has refrigerant in it, it must be recovered before the test begins. Your instructor will explain to you how to recover the refrigerant.

4. CHECK THE OIL LEVEL IN THE VACUUM PUMP AND ADD VACUUM PUMP OIL IF NEEDED.

5. After the system pressure is reduced to the correct level using the recovery unit, fasten the center gage line to the vacuum pump, Figure 3.

6. Put on your goggles and start the vacuum pump.

7. When the vacuum pump has operated long enough that the low-side compound gage has pulled down to 25 in. Hg (inches of mercury), Figure 4, open the valve to the electronic vacuum gage sensing element. Observe the compound gage as the system pressure reduces. It depends on the type of vacuum gage you have as to when it will start to indicate. Some of them do not start to indicate until around 1000 microns (1 millimeter of mercury vacuum) and will not indicate until very close to the end of the vacuum. Some gages start indicating at about 5000 microns (5 millimeters).

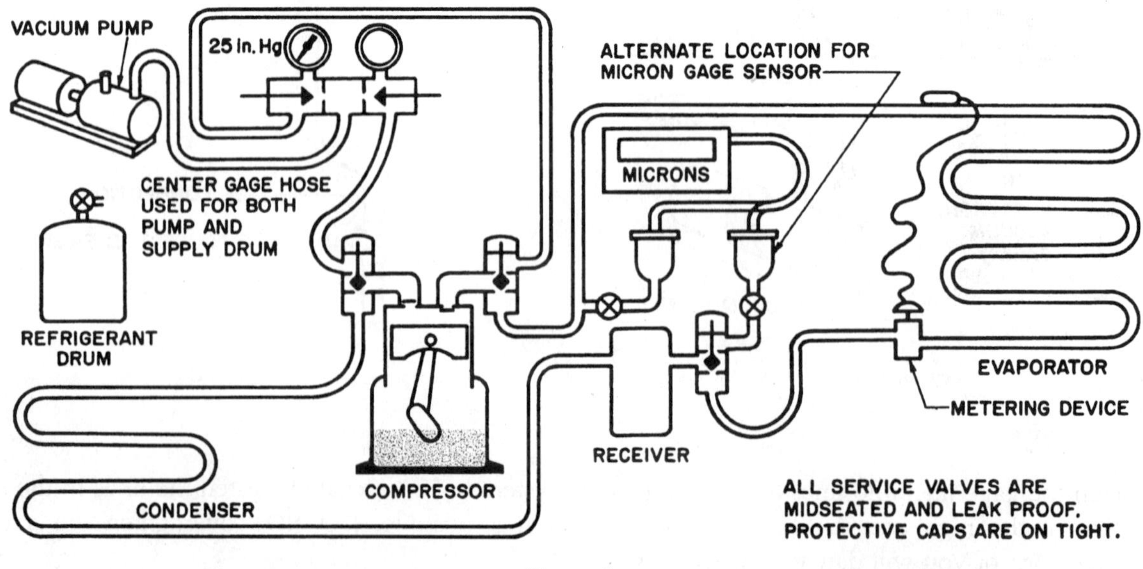

Figure 4

8. Operate the vacuum pump until the indicator reads 200 microns. NOTE: The deeper the vacuum, the slower it pulls down. It takes much more time to obtain the last few microns than the first part of the vacuum.

9. When the vacuum reaches 200 microns, you have obtained a low vacuum, one that will satisfy most any manufacturer's requirements.

10. If the system will not reach 200 microns, valve the vacuum pump off and let the system stand. If the vacuum rises, there is a leak or something is boiling out of the system. You must find the problem. A starting point is to put pressure back into the system to above atmosphere and remove the gage manifold. Evacuate the gage manifold as in a system evacuation and perform a leak test on it. It must be leak-free or you will never obtain a vacuum.

11. Upon successful completion of an evacuation, shut off the valve to the vacuum sensing element and charge enough refrigerant into the system to get the pressure above atmospheric. IT IS NEVER GOOD PRACTICE TO LEAVE A SYSTEM IN A DEEP VACUUM FOR LONG PERIODS. A LEAK COULD CONTAMINATE THE SYSTEM AND DRIERS.

MAINTENANCE OF WORK STATION AND TOOLS: Check the oil level in the vacuum pump and fill as needed with vacuum pump oil. Return the refrigeration system to the condition in which you found it. Wipe any oil off the tools and return them to their places.

SUMMARY STATEMENT: Explain in your own words how your evacuation procedure went. Don't leave out any mistakes you made.

QUESTIONS

1. Why should the electronic vacuum gage sensing element always be kept upright?

2. At what point did the electronic vacuum gage start indicating?

3. Which is the most accurate for checking a deep vacuum, an electronic vacuum gage or a compound gage?

4. What is the purpose of system evacuation?

5. What happens to a system that leaks while in a vacuum?

6. Is a vacuum the best method of removing large amounts of moisture from a system?

7. What will the results be if a fitting is leaking and the system is evacuated to a deep vacuum?

8. What is the approximate boiling pressure of refrigerant oil in a system?

9. Why is vacuum pump oil used in a vacuum pump instead of refrigerant oil?

10. How may water be released from below the oil level in a compressor crankcase?

LAB 5 Triple Evacuation

Name ______________________________ Date ______________ Grade ________

OBJECTIVES: Upon completion of this exercise, you will be able to perform a triple evacuation on a typical refrigeration system using a "U" tube mercury manometer or a gage manifold compound gage for the vacuum indicator.

INTRODUCTION: You will use a "U" tube Mercury manometer if one is available for a vacuum check. If a "U" tube manometer is not available, you may use a gage manifold and adequate time (depending on system size) to accomplish the same evacuation. You will use a refrigeration system that has service valves.

TEXT REFERENCES: Unit 5.

TOOLS AND MATERIALS: A gage manifold, gloves, goggles, high-quality vacuum pump (two-stage is preferred), "U" tube mercury manometer, refrigerant recovery system, and a cylinder of refrigerant of the same type as in the system to be evacuated.

SAFETY PRECAUTIONS: Use of caution while transferring refrigerant. Wear gloves and goggles. FOLLOW THE DIRECTIONS WITH THE MERCURY MANOMETER FOR TRANSPORTATION. DO NOT LAY THE INSTRUMENT DOWN WITHOUT THE SHIPPING PLUNGER IN PLACE OR THE MERCURY WILL BE LOST AND THE INSTRUMENT WILL BE RUINED. Check the oil in the vacuum pump. Do not begin this exercise without proper instruction in the use of the gage manifold, "U" tube manometer, refrigerant recovery system, and a vacuum pump.

PROCEDURE A

Using a "U" tube manometer.

1. Put on your goggles. Fasten the gage lines to the service valves on the low and high side of the refrigeration system.
2. Recover all refrigerant from the system down to atmospheric pressure. Your instructor will tell you how to recover the refrigerant.
3. Check the oil in the vacuum pump.
4. Fasten the "U" tube manometer to the system in such a manner that it may be valved off, Figure 1.

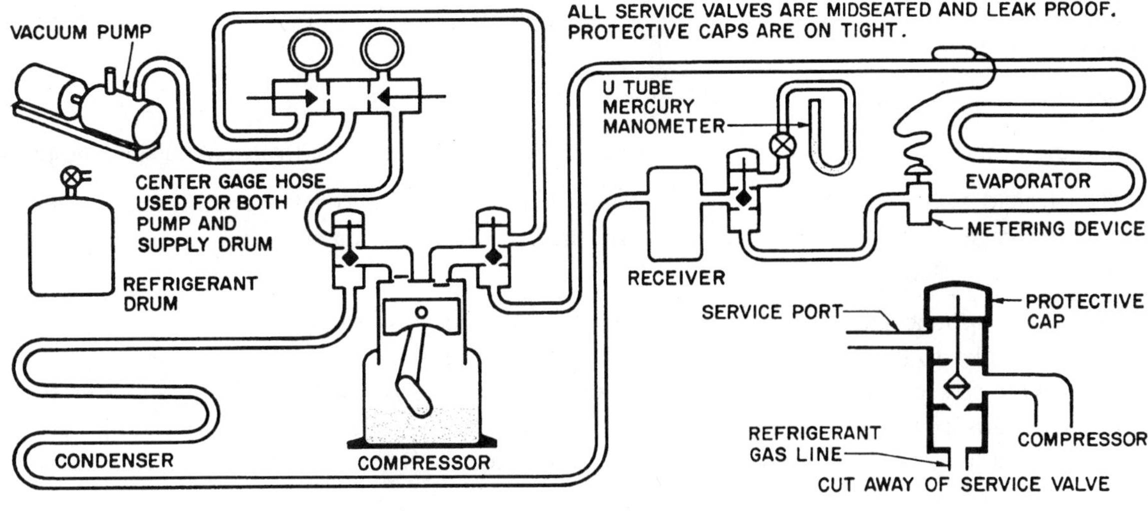

Figure 1

5. Fasten the center gage line of the gage manifold to the vacuum pump. Open the gage manifold valves and start the vacuum pump.

6. When the needle on the compound gage starts into a vacuum, you may open the valve to the mercury manometer. REMEMBER: It will not start to indicate until the vacuum is down to about 25 inches of mercury on the compound gage.

7. Allow the vacuum pump to run until the mercury columns read as in Figure 2. At this reading you can be assured that the vacuum is below 1 mm (millimeter) of mercury.

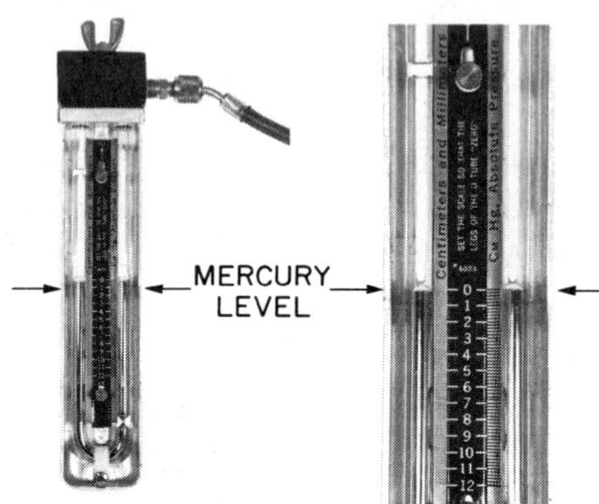

Figure 2 *Photo by Bill Johnson*

8. IMPORTANT: Valve off the mercury manometer.

9. Allow a small amount of refrigerant to enter the system, up to about 20 in. Hg on the compound gage. This is called "breaking the vacuum." Be sure to use the refrigerant that is characteristic to the system.

10. Now restart the vacuum procedure with the pump and open the valve to the mercury manometer. Allow the vacuum to reach below 1 mm Hg again and repeat steps 8 and 9.

11. When the mercury manometer has reached below 1 mm Hg again, allow the vacuum pump to run for some time, depending on the size of the system. There is no guideline for this time.

12. When some time has passed, VALVE OFF the mercury manometer and allow refrigerant to enter the system to cylinder pressure.

PROCEDURE B

Using a Gage Manifold Only.

1. Fasten gage lines to the service valves on the high and low side of the system.

2. Recover refrigerant from the system down to atmospheric pressure. Ask your instructor for the correct procedure to do this.

3. Check the oil in the vacuum pump. Add oil if it needs it.

4. Fasten the center gage line to the vacuum pump. Open the gage manifold valves and start the pump.

   NOTE: Instead of a vacuum indicator, which may not be available to every technician, we will use the compound gage reading and the audible sound the vacuum pump makes while pumping to assure that we have a deep vacuum.

5. When the compound gage has stopped moving (it will read close to 30 in. Hg vacuum), turn the gage manifold handles to the closed position. Pay attention to the tone of the vacuum pump before and after the valves are closed. If there is a tone change when closing the valves, the pump is still pumping. This is a tone comparison with the valves open and closed. NOTE: The tone of the vacuum pump may be more evident with the exhaust cap removed, Figure 3. The indicator that a deep vacuum has been obtained is

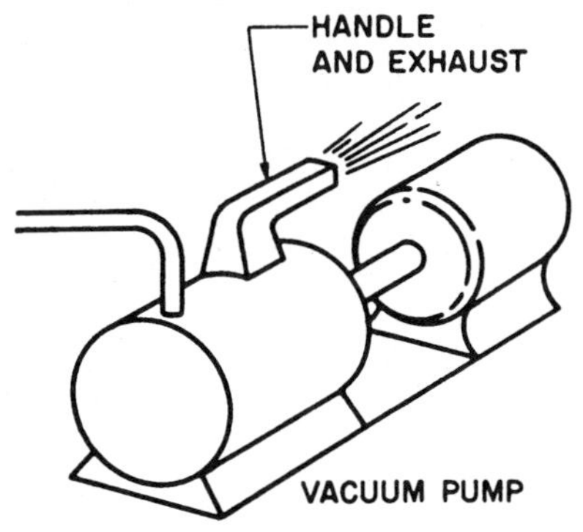

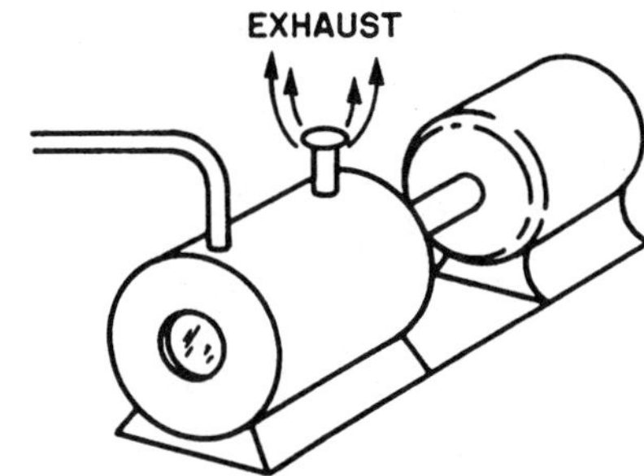

Figure 3

when there is no tone change when the valves are closed. The exhaust cap is often an oil eliminator to stop oil from leaving the pump during the early stages of the vacuum. DO NOT REMOVE IT UNTIL THE VACUUM IS BELOW 25 in. Hg. If you remove it too soon, oil will splash out of the pump with the exhaust.

6. When you are satisfied that a deep vacuum has been obtained, valve off the vacuum pump and allow a small amount of refrigerant to enter the system, up to about 20 in. Hg on the compound gage. This is called "breaking the vacuum." Be sure to use the refrigerant that is characteristic to the system.

7. Restart the vacuum pump, removing the refrigerant from the system and follow the procedure in steps 5 and 6.

8. Perform three low-vacuum evacuations on the system and then allow the refrigerant charge to fill the system to the cylinder pressure. The system is now evacuated and ready for the proper charge.

MAINTENANCE OF WORK STATION AND TOOLS: Check the oil level in the vacuum pump and add oil if needed. Clean any oil from tools and work area. Leave the refrigeration system in the manner you found it or as your instructor tells you.

SUMMARY STATEMENT: Describe what happens when a small amount of refrigerant is allowed to enter the system to "break the vacuum."

QUESTIONS

1. What is a deep vacuum?

2. Why must no foreign vapors be left in a refrigeration system?

3. What vapors does a deep vacuum pull out of the system?

4. What type of oil is used in a vacuum pump?

5. What comes out of the exhaust of a vacuum pump during the early part of an evacuation?

6. Why must vacuum pump oil only be used for a deep vacuum pump?

7. Why is a large vacuum pump required to remove moisture from a system that has been flooded with water?

LAB 6 Refrigerant Recovery

Name ________________________________ Date ______________ Grade ________

OBJECTIVES: Upon completion of this exercise, you will be able to use one of two methods for recovering refrigerant from a system.

INTRODUCTION: You will use two methods of removing refrigerant from a system. The first is to pump all available liquid into an approved refrigerant cylinder using the refrigeration system containing the refrigerant. The second is to use refrigerant recovery equipment commercially manufactured for refrigerant recovery.

TEXT REFERENCES: Unit 6.

TOOLS AND MATERIALS: Gage manifold, gloves, goggles, service valve wrench, two 8-inch adjustable wrenches, DOT- approved refrigerant cylinder, scales, refrigerant recovery system, and a refrigeration system from which refrigerant is to be recovered.

SAFETY PRECAUTIONS: DO NOT OVERFILL ANY REFRIGERANT CYLINDER, USE ONLY APPROVED CYLINDERS. Do not exceed the net weight of the cylinder. Wear gloves and goggles while transferring refrigerant.

PROCEDURES

Pumping the Liquid from the System Method

1. Fasten the low side gage line to the suction connection on the system used.
2. Fasten the high side gage line to the king valve or a liquid line connection. NOTE: All systems do not have a liquid line connection. If this system does not have one, you will need to use the second method of recovering refrigerant. Wear goggles and gloves.
3. Loosely connect the center line of the gage manifold to the refrigerant cylinder. Use only an approved recovery cylinder.
4. Bleed a small amount of refrigerant from each line through the center line and allow it to seep out at the cylinder connection. This will purge any contaminants from the gage manifold.
5. Set the cylinder on the scales and determine the maximum weight that may be allowed into the cylinder. DO NOT ALLOW THE CYLINDER WEIGHT TO EXCEED THIS MAXIMUM.
6. Start the system and open the valve arrangement between the liquid line and the cylinder. Liquid refrigerant will start to move toward the cylinder.
7. Watch the high-pressure gage and do not allow the system pressure to rise above the working pressure of the refrigerant cylinder. In most cases this is 250 psig.
8. Watch the suction pressure and do not allow it to fall below 0 psig. When it reaches 0 psig, you have removed all the refrigerant from the system that you can using this method.
9. Remove the refrigerant cylinder. You may need to recover the remaining refrigerant using recovery equipment. Recovery can be speeded up by setting the recovery cylinder in a bucket of ice, thus reducing the pressure in the cylinder.

Commercially Manufactured Recovery Unit Equipment Method

1. Read the manufacturer's directions closely. Some units recover refrigerants only and some recover and filter the refrigerant.
2. Connect the recovery unit to the system, following the manufacturer's directions for the particular type of system you have.
3. Operate the recovery system as per the manufacturer's directions.
4. Upon completion of the operation, the refrigerant will be recovered in an approved container.

MAINTENANCE OF WORK STATION AND TOOLS: Return the system to the condition your instructor requests you to. Return all tools and materials to their proper places.

SUMMARY STATEMENT: Describe the difference in recovery only and recycling.

QUESTIONS

1. After a hermetic motor burn, what contaminant might be found in the recovered refrigerant?
2. What would the result be if a recovery cylinder were filled with liquid and allowed to warm up?
3. Why must liquid refrigerant be present for the pressure-temperature relationship to apply?
4. If an air conditioning unit using R-22 were located totally outside and the temperature was 85°F, what would the pressure in the system be with the unit off? (Hint: It will follow the P and T relationship.)
5. When an evaporator using R-12 is boiling the refrigerant at 21°F, what will the suction pressure be? __________ psig
6. When an evaporator using R-22 is boiling the refrigerant at 40°F, what will the suction pressure be? __________ psig
7. When the suction pressure of an R-22 system is 60 psig, at what temperature is the refrigerant boiling? __________°F
8. When the suction pressure of an R-502 system is 20 psig, at what temperature is the refrigerant boiling? __________°F
9. What is the condensing temperature for a head pressure of 175 psig for an R-12 system? __________°F
10. What is the condensing temperature for a head pressure of 296 psig for an R-22 system? __________°F

LAB 7 Charging a System

Name ______________________________ Date ______________ Grade ________

OBJECTIVES: Upon completion of this exercise, you will be able to add refrigerant to a charging cylinder and charge refrigerant into a typical refrigeration system using a charging cylinder.

INTRODUCTION: You will apply gages to a typical capillary tube refrigeration system, recover the existing charge and evacuate the system to a deep vacuum. You will then add enough charge to a charging cylinder, connect it to the system and measure in the correct charge. You will need to use a package system or one with a known charge, and one with service valves or Schrader valve ports.

TEXT REFERENCES: Unit 7.

TOOLS AND MATERIALS: A vacuum pump, gage manifold, refrigerant recovery equipment, goggles, gloves, refrigerant that is characteristic to the system, graduated charging cylinder, straight blade screwdriver, 1/4" and 5/16" nut drivers, and a capillary tube refrigeration system as described above.

SAFETY PRECAUTIONS: Wear gloves and goggles while transferring refrigerant to and from a system. DO NOT overcharge the charging cylinder. Follow all instructions carefully. Your instructor must give you instruction in using the gage manifold, refrigerant recovery equipment, adding refrigerant to the charging cylinder and to the system, using a vacuum pump and evacuating a system before you proceed with this exercise.

PROCEDURES

1. Connect the refrigerant recovery equipment to the system and recover all refrigerant from the system to be charged. Your instructor will instruct you as to where and how to recover the refrigerant.
2. Connect the gage lines to the service valves or Schrader valve ports on the high and low sides of the system.
3. Check the oil level in the vacuum pump, add oil if needed, connect it to the center line of the manifold and start the pump. Open the gage manifold valves.
4. Determine the correct charge for the system and write it here for reference. __________ ounces
5. While the evacuation is being performed, connect the charging cylinder to the cylinder of refrigerant, Figure 1.

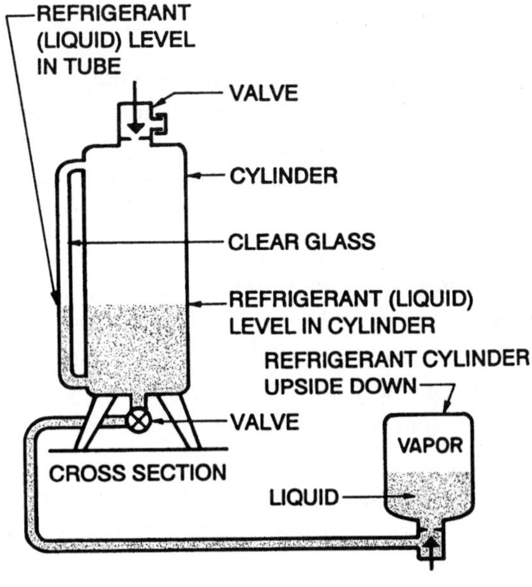

Figure 1

6. When the charging cylinder has enough refrigerant to charge the system and the vacuum has reached the correct level (deep vacuum is determined by sound or instrument), you are ready to charge the refrigerant into the system.

7. Shut off the vacuum pump and connect the charging cylinder as shown in Figure 2.

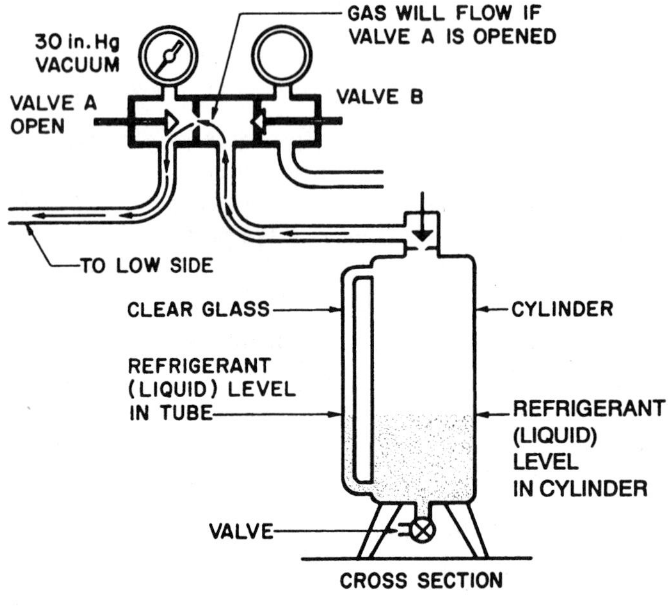

Figure 2

8. Purge a small amount of refrigerant from the gage manifold line by loosening the center gage line at the manifold after the connection is made because some air will be drawn into the gage line when disconnecting from the vacuum pump.

9. Turn the outside cylinder on the charging cylinder to correspond to the pressure in the cylinder. For example, if the pressure gage at the top of the cylinder reads 75 psig (this would be R-12), turn the dial to the 75 psig scale and read the exact number of ounces on the sight glass of the charging cylinder, Figure 3. Record it here. ________ ounces

10. Subtract the amount of the unit charge from the charging cylinder level and record it here. _________ ounces This is the stopping point of the change.

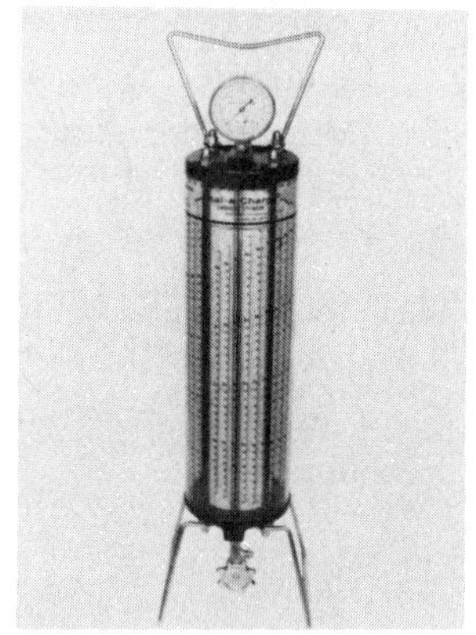

Figure 3 *Photo by Bill Johnson*

11. Open the vapor valve at the top of the cylinder and the low-side valve on the gage manifold. The charge will begin to fill the system.

12. When the system is above 0 psig, disconnect the high-side gage line. The reason for this is that the charging cylinder pressure will soon equalize to the same pressure as the system and the refrigerant will stop flowing. You may have to start the system to pull in the correct charge. You do not want any condensed refrigerant in the high-side gage line. Some charging cylinders have a heater in the base to keep the pressure up and the system may not have to be started.

13. Keep adding refrigerant vapor until the correct low level appears on the sight glass on the charging cylinder (Figure 4) and then shut off the suction valve on the gage manifold.

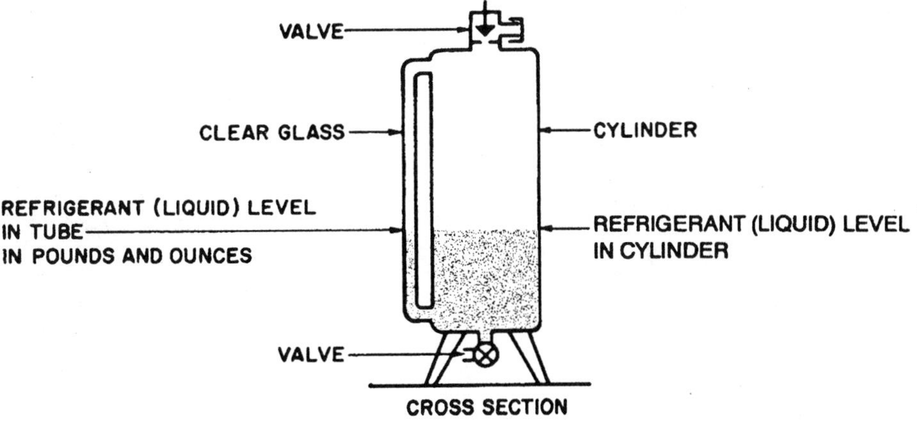

Figure 4

MAINTENANCE OF WORK STATION AND TOOLS: Check the vacuum pump oil and add if needed. Wipe up any oil in the work area. Return all tools to their proper places. Replace panels on the unit.

SUMMARY STATEMENT: Describe the calculations you had to make to determine and make the correct charge for the system.

QUESTIONS

1. What are the graduations on the charging cylinder?

2. Why is it necessary to turn the dial to the correct pressure reading and use this figure for the charge?

3. What scale or graduation was the charge expressed in on the system you charged, pounds or pounds and ounces?

4. How many ounces in a pound?

5. How many ounces in one-fourth pound?

6. Why was the high-side line disconnected before starting the unit?

7. How may refrigerant be charged into the system if the cylinder pressure drops?

8. How do some charging cylinders keep the pressure from dropping as the charge is added?

9. What type of metering device normally requires an exact charge?

10. Do all systems have sight glasses in the liquid line to aid in charging?

LAB 8 Charging a System with Refrigerant Using Scales

Name ______________________ Date ______________ Grade __________

OBJECTIVES: Upon completion of this exercise, you will be able to charge a typical refrigeration system using scales to weigh the refrigerant.

INTRODUCTION: You will recover refrigerant, if any, from a refrigeration system, evacuate the system to prepare it for charging using refrigerant weight, then charge the system to the correct operating charge.

TEXT REFERENCES: Unit 7.

TOOLS AND MATERIALS: A gage manifold, gloves, goggles, refrigerant recovery equipment, a high-quality vacuum pump, a set of accurate scales (electronic scales are preferred because of their ability to count backwards, but the exercise will be written around standard platform scales), and a cylinder of the proper refrigerant for the system to be charged. You will need a refrigeration system that has service valves or ports and that has a known required charge.

SAFETY PRECAUTIONS: Wear gloves and goggles any time you are transferring refrigerant. Your instructor must give you instruction in recovering and transferring refrigerant before starting this exercise.

PROCEDURES

1. Recover any refrigerant that may be in the system down to atmospheric pressure. Your instructor will tell you how to recover the refrigerant correctly.

2. While wearing goggles, connect the gages to the high- and the low-side service valves or Schrader ports.

3. Add oil to the vacuum pump if it is not at the proper level. When the level is correct, connect it to the center gage line and turn it on.

4. Determine the correct charge for the system from the nameplate or the manufacturer's specifications and record it here. __________ ounces.

5. Set the refrigerant cylinder on the scales with the vapor line up. Make a provision to secure the gage line when it is disconnected from the vacuum pump and connected to the cylinder on the scales so it will not move and vary the reading. A long gage line is preferred for this, Figure 1.

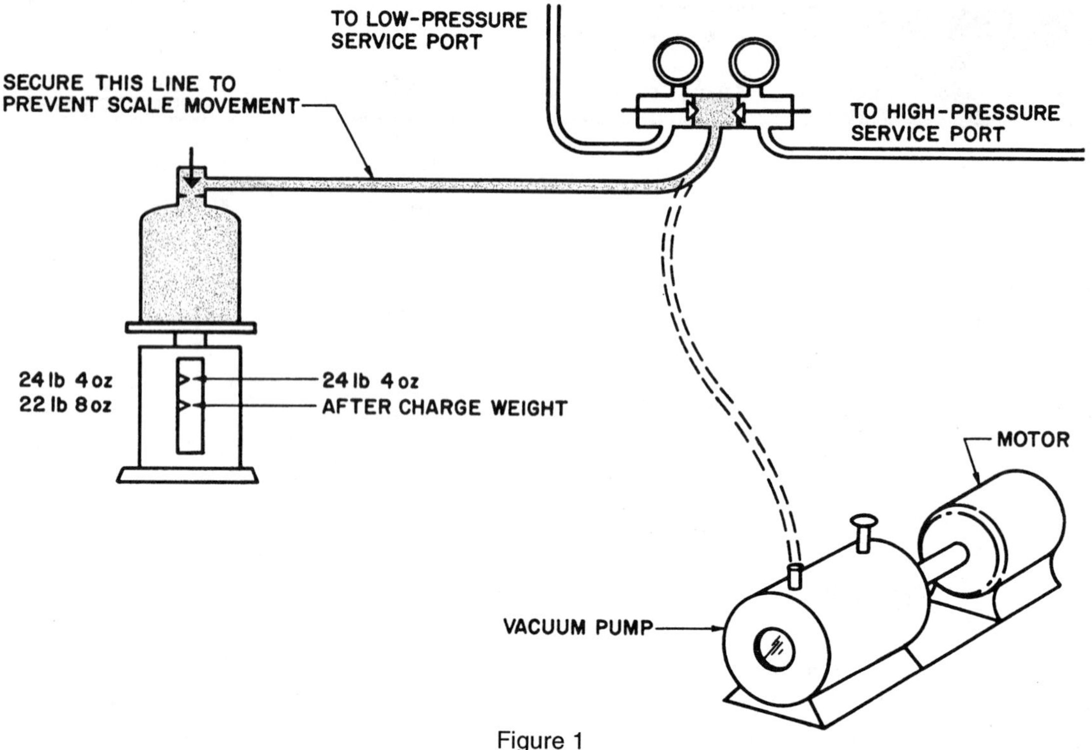

Figure 1

6. When the vacuum pump has evacuated the system to the correct point (the sound test mentioned in Lab 5 may be used if no vacuum gage is available), close the gage manifold valves, turn off the vacuum pump, disconnect the center gage line from the vacuum pump and connect it to the refrigerant cylinder. Don't forget to secure it.

7. Allow a very small amount of refrigerant vapor to bleed out the center gage line at the manifold because when you disconnect the vacuum pump, vapor will be drawn in.

8. Record the cylinder weight in ounces here. __________ ounces

   NOTE: The scales may read in pounds and tenths of a pound. Convert this to ounces to make calculations easier. For instance, 0.4 of a lb. equals 6.4 ounces (16 oz × 0.4 = 6.4).

9. Calculate what the reading on the scale should be when the system has been completely charged and enter here. __________ ounces

10. Open the low-side gage manifold valve and allow refrigerant vapor to start entering the system.

11. When the system pressure is above atmospheric, disconnect the high-side line. The reason for this is because refrigerant will condense in this high-side line when the system is started and the charge will not be correct.

12. If all of the charge will not enter the system, the system may be started to lower the system pressure so that the rest of the charge will be pushed in.

13. When the complete charge is in the system according to the scales, disconnect the low-side line and run the system if it is ready to run.

MAINTENANCE OF WORK STATION AND TOOLS. Wipe any oil off the tools and work area. Check the oil in the vacuum pump and fill with approved vacuum pump oil if needed. Replace any panels on the unit. Place all tools in their respective places.

SUMMARY STATEMENTS: Describe the procedure for determining the amount of refrigerant to be charged into the system. Describe the calculations you made regarding the use of the scales. Include any conversions from pounds to ounces.

QUESTIONS

1. Why is the high-side line disconnected before the system is started?

2. Why is the line to the refrigerant cylinder purged after evacuation and before charging?

3. Why is the gage line secured when the refrigerant cylinder is on the scales while transferring refrigerant?

4. Why would the cylinder pressure drop when charging vapor into a system?

5. What can be done to keep refrigerant moving into the system if the cylinder pressure drops and the cylinder must remain on the scales?

6. Why is vapor only recommended for a charging system?

7. What would happen if liquid refrigerant were to enter the compressor cylinder while it was running?

8. What type of oil is recommended for a high-vacuum pump?

9. How many ounces in a pound?

10. How many ounces in three-tenths (0.3) of a pound?

LAB 9 Making Voltage and Amperage Readings with a VOM

Name ______________________ Date ____________ Grade ________

OBJECTIVES: Upon completion of this exercise, you will be able to make voltage and amperage readings on actual operating equipment using a VOM. You will be able to do this under the supervision of your instructor.

INTRODUCTION: You will be using a VOM to make voltage readings and a clamp-on-type ammeter to make AC amperage readings. Your ammeter may be an attachment to your VOM or it may be designed to use independently.

TEXT REFERENCES: Unit 8.

TOOLS AND MATERIALS: A VOM with insulated alligator clip test leads, a clamp-on-type ammeter, 1/4" and 5/16" nut drivers, and a straight blade screwdriver.

SAFETY PRECAUTIONS: Working around live electricity can be very hazardous. Use a meter with insulated alligator clips on the ends of the leads. Make all meter connections with the POWER OFF. Have your instructor inspect and approve all connections before turning the power on. Your instructor should have given you thorough instruction in using the VOM and clamp-on ammeter. The meters will vary in design and you will need specific instruction in the use of each. The scales on your meter may differ from those indicated in this exercise.

PROCEDURES

Making Voltage Readings

1. With the power off, locate the control voltage transformer in an air conditioning or heating unit.
2. Determine the output (24 V) terminals.
3. Turn the function switch to AC and the range switch to 50 V on your VOM. The black test lead should be in the common (–) jack and the red test lead in the (+) jack in your meter.
4. Connect the insulated alligator clips to the transformer output (24 V) terminals.
5. Have your instructor check the connections. Turn the power on. (You may have to adjust the thermostat to call for heating or cooling.) Record the AC reading: __________
6. Turn off the power.
7. Remove the leads.
8. You will now measure the voltage at the input side of the transformer. The meter function switch should be set at AC. Set the range switch at 250 V.
9. With the power off, fasten the meter test leads with alligator clips to the two terminals at the input side of the control transformer.
10. Ask your instructor to check the connections. With approval turn the power on. Record the AC voltage: __________

11. Turn the power off.
12. Remove the leads.
13. Ask your instructor to show you the location of the 230-V terminals to the compressor of an air conditioning unit.
14. Be sure the power to the compressor is turned off.
15. Set the meter function switch at AC and the range switch at 500 V. (Always select a setting higher than the anticipated voltage.)
16. Connect the meter lead alligator clips to the terminals.
17. Have your instructor inspect the connections and with approval turn the power on. Record the voltage reading: __________
18. Turn power off.
19. Remove the leads.

Making Amperage Readings

1. With the power off, place the clamp-on ammeter jaws around a wire to the fan motor.
2. Have your instructor check your setup, including any necessary settings on your meter.
3. With approval turn the power on. Record the amperage reading: __________
4. Turn the power off.
5. Remove the meter.
6. Have your instructor show you a wire leading to an air conditioning compressor or another component that has a 230-V source.
7. With the power off, place the ammeter jaws around this wire.
8. Have your instructor inspect your meter and with approval turn the power on. Record the amperage reading: __________
9. Turn off the power.
10. Remove the meter.
11. With the power off, disconnect one lead from the output of the 24-V control transformer. Wrap a small wire ten times around one jaw of the ammeter. Connect one end of this wire to the terminal where you just removed the wire. Connect the other end to the wire you removed.
12. Ask your instructor to check your setup including any settings you may need to make on your meter. With the instructor's approval turn the power on. Record your amperage reading: __________ Your reading will actually be ten times the actual amperage because of the ten loops of wire around the jaws of the ammeter. Divide your reading by ten and record the actual amperage: __________
13. Turn off the power. Remove the wire and ammeter. Connect the wire back to the transformer terminal as you originally found it.

MAINTENANCE OF WORK STATION AND TOOLS: Replace all panels on equipment with correct fasteners. Return all meters and tools to their proper places.

SUMMARY STATEMENTS: Describe in your own words how you would make a voltage reading in the 230-V range.

Describe how you would use a clamp-on ammeter to make an AC amperage reading.

QUESTIONS

1. Why is alternating current rather than direct current normally used in most installations?

2. Describe why the voltage is different from the input to the output of a transformer.

3. Is a typical control transformer a step-up or step-down transformer?

4. Describe a VOM.

5. What are typical settings for the function switch on a VOM?

6. What is the range switch used for on a VOM?

7. Explain how the magnetic field can be increased around an iron core.

8. Explain how a solenoid switch works.

9. Describe inductance.

10. When measuring voltage, is the meter connected in series or parallel?

LAB 10 Measuring Resistance and Using Ohm's Law

Name ________________________ Date ______________ Grade ________

OBJECTIVES: Upon completion of this exercise, you will be able to make current, voltage, and resistance readings. You will also determine the current, voltage and resistance of a circuit using Ohm's Law.

INTRODUCTION: You will be using a VOM to make resistance readings and a VOM to check your calculations using Ohm's Law.

TEXT REFERENCES: Unit 8.

TOOLS AND MATERIALS: A VOM with insulated alligator clip test leads, a clamp-on-type ammeter, 1/4" and 5/16" nut drivers, and a straightblade screwdriver.

SAFETY PRECAUTIONS: Use a meter with insulated alligator clips on the ends of the leads. Make all meter connections with the power off. Have your instructor inspect all connections before turning the power on. Your instructor should have given you thorough instruction in using all features of your meter.

PROCEDURES

1. Make the zero ohms adjustment on the VOM in the following manner: Set the function switch to either –DC or +DC. Turn the range switch to the desired ohms range. Probably R × 100 will be satisfactory for this exercise. Connect the two test leads together and rotate the zero ohms control until the pointer indicates zero ohms.
2. With the power off, disconnect one lead to an electric furnace heating element.
3. Clip one lead of the meter to each terminal of the element. Do not turn the power on. Read and record the ohms resistance. __________ Disconnect the meter leads.
4. Connect the heating element back into its original circuit. Place the clamp-on ammeter jaws around one lead to the heating element.
5. Have your instructor inspect and approve your setup. Turn the power on and quickly read the amperage. Record the amperage. __________ You must take the reading quickly because as the element heats up the resistance increases. Turn the power off.
6. Turn the function switch on the VOM meter to AC and set the range switch to 500 V.
7. With the power off, connect a meter lead to each terminal on the heating element.
8. Have your instructor inspect and approve your setup. Turn the power on. Record the voltage: __________ Turn the power off.
9. Check your readings by using Ohm's Law. To check the amperage reading use the following formula: $I = E/R$
10. Substitute the voltage and resistance reading from your notes. Divide the voltage by the resistance. Your answer should be close to your amperage reading.

MAINTENANCE OF WORK STATION AND TOOLS: Remove test leads from your meter and coil them neatly or follow your instructor's directions. Replace any panels. Return all tools and equipment to their proper place.

SUMMARY STATEMENTS: Describe the process of zeroing the ohms on a meter and measuring the resistance of a component.

QUESTIONS

1. Describe the difference between a step-up and step-down transformer.

2. What do the letters VOM stand for? What does a typical VOM measure?

3. Describe how to make the zero ohms adjustment on a VOM.

4. What is inductance?

5. Describe how you would use a typical clamp-on ammeter to measure the amperes in a circuit.

6. Explain how a larger diameter wire can safely carry more current than a smaller size.

7. What can happen if a wire too small in diameter is used for a particular installation?

8. How can circuits be protected from current overloads?

9. What are two methods used in most circuit breakers to protect electrical circuits?

10. What is a ground fault circuit interrupter?

LAB 11 Automatic Controls

Name ______________________ Date ______________ Grade ________

OBJECTIVES: Upon completion of this exercise, you will be able to identify several temperature-sensing elements and state why they respond to temperature changes.

INTRODUCTION: You will use several types of temperature-sensing devices, subjecting them to temperature changes and recording their responses.

TEXT REFERENCES: Unit 9.

TOOLS AND MATERIALS: A straight blade and a Phillips screwdriver, a millivolt meter if available (a digital meter should read down to the millivolt range), a VOM for measuring ohms, a source of low heat (the sun shining in an area or heat from a warm air furnace), an air acetylene unit, ice and water, a fan-limit control, a remote-bulb thermostat (close on a rise in temperature refrigeration type) with a temperature range close to room temperature, a thermocouple from a gas furnace, and a thermistor temperature tester.

SAFETY PRECAUTIONS: Care should be used with the open flame. DO NOT APPLY THE FLAME HEAT TO THE REMOTE-BULB THERMOSTAT.

PROCEDURES

Checking Bimetal and Remote-Bulb Devices

1. Lightly fasten the fan limit control straight up in a vise.
2. Connect the two ohmmeter leads together, turn the meter to the R × 1 scale and zero the meter. If the meter is a digital meter, it is not necessary to do the zero adjust.
3. Remove the cover of the fan limit control and fasten the ohmmeter leads to the terminals for the fan operation. Your instructor may need to show you which terminals, Figure 1. The meter should read infinity. This indicates an open circuit and is normal with the control at room temperature.

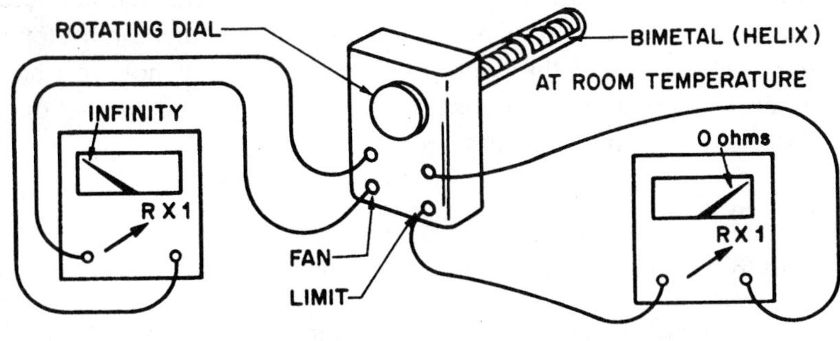

Figure 1

4. Light the air acetylene torch and pass the flame close to, but not touching, the fan control bimetal element. Pass the flame back and forth. Notice the dial on the front of the control start to turn as the element is heated. Soon the control contacts will close and the ohmmeter will read 0 ohms resistance. The control may give an audible click at this time.
5. Remove the heat source as soon as the meter reads 0 ohm and allow the control to cool. In a few moments, the control contacts will open and the ohmmeter will again show an open circuit by reading infinity. This closing and opening of these contacts would start and stop a fan if in a furnace circuit.

6. Fasten the ohmmeter leads to the terminals that control the high-limit action. Again, your instructor may need to help you. The meter should show 0 ohms resistance. This circuit is normally closed at room temperature.

7. Again move the torch slowly back and forth close to the sensing element. You may hear the audible click as the contacts close to start the fan. This is normal. With a little more heat applied, you should then hear the audible click of the high-limit contacts opening to stop the heat source. This will cause the meter to read infinity, indicating an open circuit.

8. Allow the control to cool to room temperature. The limit contacts will close first, then the fan contacts will open.

9. Remove the cover from a remote-bulb thermostat.

10. Fasten the leads from the ohmmeter to the terminals on the remote-bulb thermostat, Figure 2.

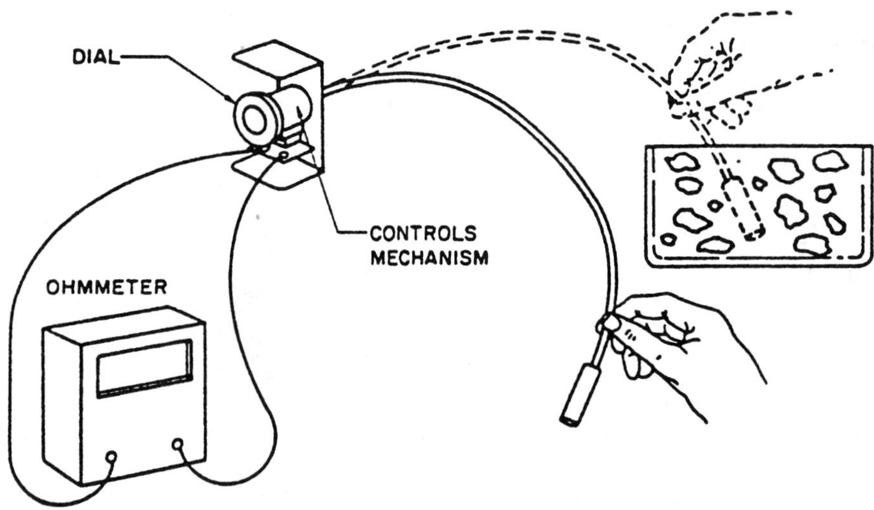

Figure 2. Checking the action of a remote-bulb thermostat by first holding it in your hand and then placing it in ice and water.

If the ohmmeter shows that the circuit is open, turn the adjustment dial until the contacts just barely close.

11. Position the control so that you can see the action inside as the temperature begins to move its mechanism. Now grasp the control bulb in your hand and observe the control levers start to move. Soon you should hear the audible click of the control contacts closing. If not, you have the adjustment dial turned so high that hand temperature will not cause the control to function. Let the bulb cool and change the adjustment dial downward and start again.

12. Now set the control at 40°F.

13. Place the control bulb in the ice and water mixture and observe the contacts and their action.

14. When the contacts open, remove the bulb from the water so that it will be exposed to normal room temperature. The contacts should close again in a few minutes.

Checking a Thermistor Thermocouple Temperature Device

15. Leave the thermistor device at room temperature until it has stabilized. Record the temperature. __________°F

16. Place the lead in the palm of your hand and hold it tightly until it no longer changes. Record the temperature. __________°F

17. Place the lead in an ice and water mixture and stir it until it stops changing. Record the temperature. __________°F

18. Place the lead next to a heat source, not greater than the range of the meter, such as close to a light bulb or in the sun. Record the temperature. __________°F

19. Fasten a thermocouple (such as a gas furnace thermocouple) lightly in a vise. Fasten the millivolt meter leads to the end of the thermocouple, Figure 3.

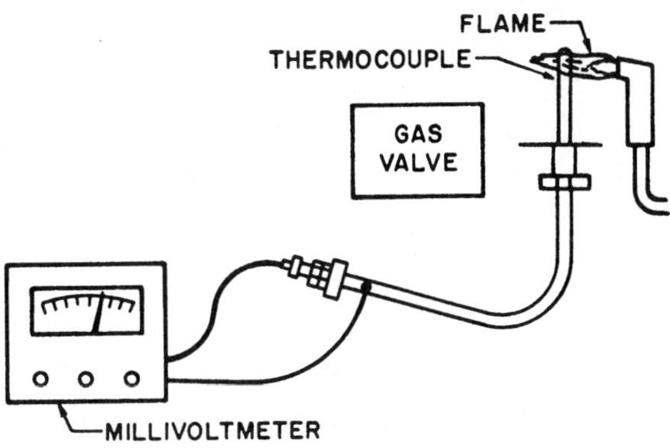

Figure 3

20. Light the torch and pass the flame lightly over the end of the thermocouple. If the meter needle moves below 0 volts, reverse the leads. This is direct current and the meter may be connected wrong. DO NOT OVERHEAT THE THERMOCOUPLE LEAD. Record the thermocouple voltage. __________mV

21. Allow the thermocouple to cool and watch the meter needle move towards 0 millivolts.

MAINTENANCE OF WORK STATION AND TOOLS: Turn off the torch at the tank valve. Bleed the gage. Return all tools, materials, and equipment to their respective places.

SUMMARY STATEMENTS: Describe the action of the four temperature-sensitive devices that you checked and observed. Tell how each functions with a temperature change.

A. Bimetal device:

B. Liquid-filled bulb device:

C. Thermocouple device:

D. Thermistor device:

QUESTIONS

1. State three methods used to extend the length of a bimetal device for more accuracy.

2. How does the action of a room thermostat make and break an electrical circuit to control temperature?

3. What causes a bimetal element to change with a temperature change?

4. Is a bimetal sensing element used to control liquids by submerging it in the liquid?

5. Name two applications for a bimetal sensing element.

6. What causes a mechanical action in a liquid-filled bulb with a temperature change?

7. Name an application for a liquid-filled bulb sensing device.

8. What is the tube connecting the bulb to the control called on a liquid-filled remote-bulb control?

9. Describe how a thermocouple reacts to temperature change.

LAB 12 Controlling Temperature

Name ______________________________ Date ____________ Grade ______

OBJECTIVES: Upon completion of this exercise, you will be able to use a temperature-measuring device to determine the accuracy of thermostats while observing the control of the temperature of the space.

INTRODUCTION: You will use a thermometer with its element located at a temperature-control device to determine the accuracy of a thermostat used to control space temperature.

TEXT REFERENCES: Unit 10.

TOOLS AND MATERIALS: A quality thermometer (electronic type with a thermistor or thermocouple is preferred because of its response time), a VOM, a small screwdriver, an air conditioning system with supply and return duct (a classroom system will do well) and a thermostat used in a typical room.

SAFETY PRECAUTIONS: You will be using the screwdriver to make corrections in the temperature control setting. Remove the control's cover prior to starting the test. Your instructor must give you instruction in operating the particular system and room thermostat before you begin.

PROCEDURES

Checking Product Temperature Thermostat

1. Locate the remote bulb for the temperature control in the return air for the system, Figure 1.

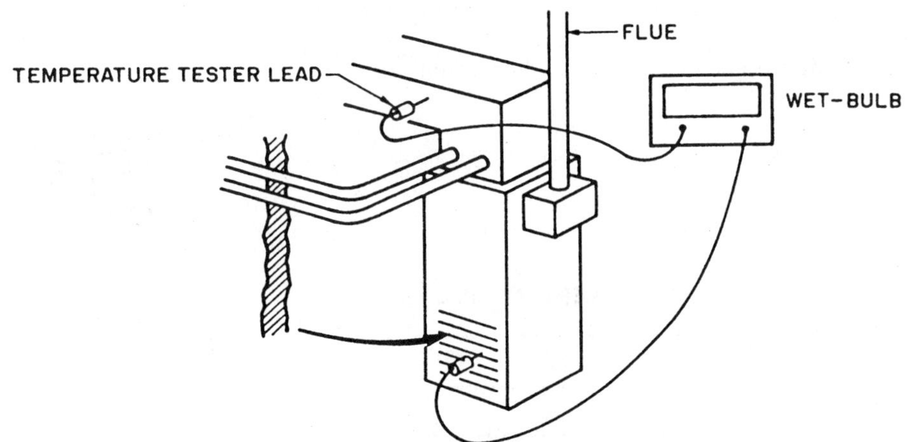

Figure 1 Location of temperature tester sensors. SHUT THE DOOR IF THE CASE IS NOT AN OPEN CASE.

2. Place a second sensor in the supply air for the system.

3. After 5 minutes, record the following. Probe A, in the return air __________°F. Probe B, in the supply air __________°F.

4. Listen for the compressor to stop and record the same 2 temperatures when it stops.

A. ________°F B. ________°F

Record every 5 minutes:

A. ________°F B. ________°F

A. ________°F B. ________°F

A. ________°F B. ________°F

A. ________°F B. ________°F

A. ________°F B. ________°F

A. ________°F B. ________°F

A. ________°F B. ________°F

5. Set the control for 5 degrees lower.

Record every 5 minutes:

A. ________°F B. ________°F

A. ________°F B. ________°F

A. ________°F B. ________°F

A. ________°F B. ________°F

A. ________°F B. ________°F

A. ________°F B. ________°F

A. ________°F B. ________°F

Checking a Space Temperature Thermostat

6. Remove the cover of a room thermostat so the mercury contacts may be observed.
7. Place the lead of a thermometer next to the thermostat, at the same level as the bimetal element, Figure 2.

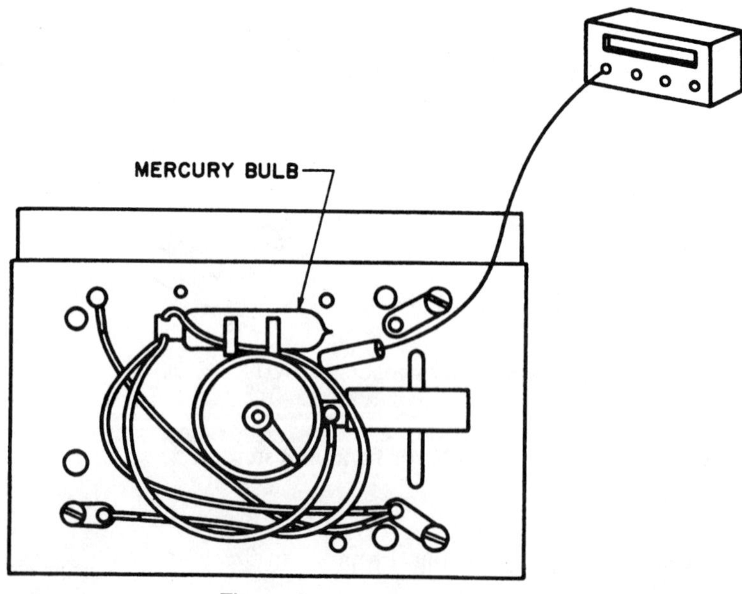

Figure 2

8. Set the VOM for measuring resistance using R × 1 scale. Switch the thermostat to the cooling mode and place one meter lead on "R" and the other on "Y." Set the thermostat so that the meter shows an open circuit. When the space around the thermostat warms, the contacts will close. Stand back from the thermostat (your body heat will affect its response) and observe the action of the contacts. If the room temperature is varying, the contacts should open and close from time to time. If they do not function in about 5 minutes, you may move close to the thermostat so your body heat will cause a response. This can be verified by the use of an ohmmeter if the thermostat is on the workbench.

9. Record the time and temperature changes.

Time ________ Temperature ________°F

Time ________ Temperature ________°F

Time ________ Temperature ________°F

Time ________ Temperature ________°F

Time ________ Temperature ________°F

Time ________ Temperature ________°F

Time ________ Temperature ________°F

Time ________ Temperature ________°F

MAINTENANCE OF WORK STATION AND TOOLS: Clean the work station and return all equipment to its respective place.

SUMMARY STATEMENT: Describe the action of the two thermostats in relation to the thermometer that you used. Did the thermostats need calibrating? Describe how this calibrating was done or could have been done.

QUESTIONS

1. Which type of thermometer would respond the fastest? (a mercury bulb or an electronic thermistor)

2. How does a bimetal sensing element make and break an electrical circuit?

3. What type of action does a bimetal element produce with a temperature change?

4. What type of action does a thermocouple produce with a change in temperature?

5. What type of action does a thermistor produce with a change in temperature?

6. Which of the following, a thermistor or a bimetal sensing element, is used more commonly in room thermostats?

7. Why does the product temperature lag behind the air temperature in a refrigerated box?

8. Explain the differential setting on a temperature control.

9. Describe a good location for a space temperature thermostat?

10. Describe the action of a typical bimetal thermostat.

LAB 13 Pressure Sensing Devices

Name ______________________ Date ______________ Grade ________

OBJECTIVES: Upon completion of this exercise, you will be able to determine the setting of a refrigerant-type, low-pressure control and reset it for low-charge protection.

INTRODUCTION: You will use dry nitrogen as a refrigerant and determine the existing setting of a low-pressure control, set that control to protect the system from operating under a low-charge condition, and use an ohmmeter to prove the control will open and close the circuit.

TEXT REFERENCES: Unit 13.

TOOLS AND MATERIALS: An adjustable type of low-pressure control, such as in Figure 1, screwdriver, 1/4" flare union, goggles, gloves, ohmmeter, gage manifold and a cylinder of dry nitrogen.

SAFETY PRECAUTIONS: You will be working with dry nitrogen under pressure, and care must be taken while fastening and loosening connections. Wear goggles and gloves.

Figure 1 Typical low-pressure control. *Photo by Bill Johnson*

PROCEDURES

1. Connect the low-pressure control to the gage manifold as shown in Figure 2.

2. Typically for an R–22 system, the lowest allowable pressure for an air conditioning system is 40 psig. A typical high setting may be 80 psig. Set the control at these approximate settings.

3. Turn both valves to the off position in the gage manifold.

4. Turn the cylinder valve on, allowing pressure up to 100 psig to enter the center line of the gage manifold.

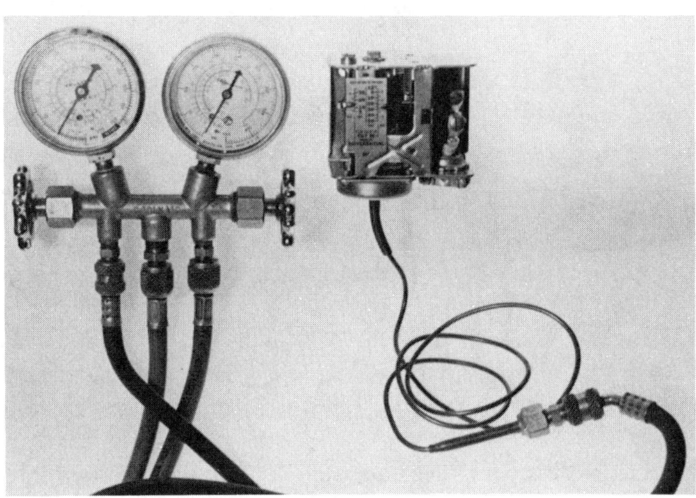

Figure 2 *Photo by Bill Johnson*

5. Attach the ohmmeter leads to the terminals on the low-pressure control. The meter should read infinity, showing that the control contacts are open. If the contacts are closed, the control is set for operation below atmosphere and must be reset for a higher value.

6. Slowly open the gage manifold valve to the low-side gage line and record the pressure at which the control contacts close. NOTE: IF THE CONTROL MAKES SO FAST THAT YOU CAN'T FOLLOW THE ACTION, ALLOW A SMALL AMOUNT OF GAS TO BLEED OUT AROUND THE GAGE LINE AT THE FLARE UNION WHILE SLOWLY OPENING THE VALVE. You may have to repeat this step several times in order to determine the exact pressure at which the control contacts close. Record the pressure reading at which the contacts close.__________ psig (This is the "cut-in" pressure of the control.)

7. With the control contacts closed and pressure on the control line, shut the gage manifold valves. Pressure is now trapped in the gage line and control.

8. Slowly bleed the pressure out of the control by slightly loosening the connection at the flare union. Record the pressure at which the control contacts open.__________ psig (This is the "cut out" pressure of the control.) You may have to perform this operation several times to get the feel of the operation. YOU NOW KNOW WHAT THE CONTROL IS SET AT TO CUT IN AND OUT. Record here.
   - Cut in __________ psig
   - Cut-out __________ psig
   - Differential (cut-in minus cut-out)

     __________ psig

Procedures for Resetting the Pressure Control

9. Determine the setting you want to use and record here.
   - Cut-in __________ psig
   - Cut-out __________ psig
   - Differential __________ psig

10. Set the control range indicator at the desired cut-in point, Figure 3.

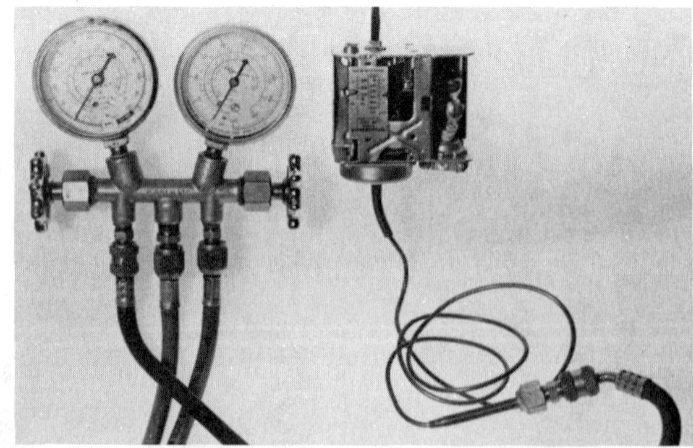

Figure 3 Setting the range adjustment. *Photo by Bill Johnson*

11. Set the control's differential indicator to the desired differential, Figure 4.

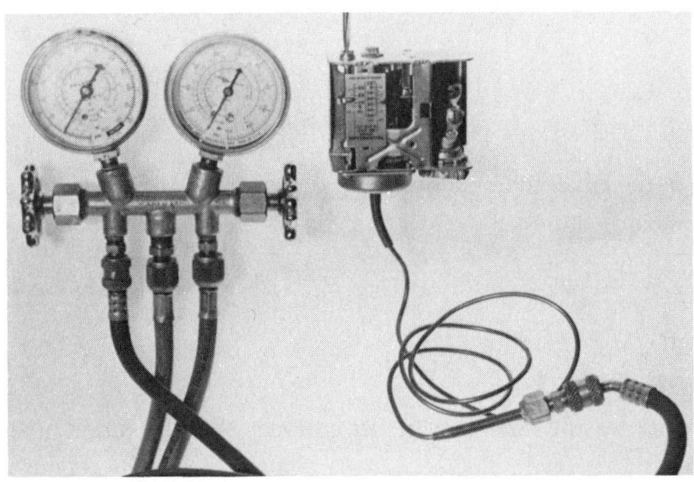

Figure 4 Setting the differential adjustment. *Photo by Bill Johnson*

12. Recheck the control pressure settings. Record here.

    - Cut-in __________ psig
    - Cut-out __________ psig
    - Differential __________ psig

    You may notice that the pointer and the actual cut-in and differential are not the same. The pointer is only an indicator. The only way to properly set a control is to do just as you have done, then make changes until the control responds as desired. If the control differential is not correct, make this adjustment first. For more differential, adjust the control for a higher value. For example, if the differential is 20 psig and you want 40 psig, turn the differential dial to a higher value.

13. If the control needs further setting, make the changes and recheck the control using pressure. Record the new values here.

    - Cut-in __________ psig
    - Cut-out __________ psig
    - Differential __________ psig

MAINTENANCE OF WORK STATION AND TOOLS: Return all tools and materials to their respective places. Clean your work area.

SUMMARY STATEMENT: Describe the range and differential functions of this control.

QUESTIONS

1. Why should a unit have low-pressure protection when a low charge is experienced?

2. What typical pressure sensing device may be used in a low-pressure control?

3. What is the function of the small tube connecting the pressure-control sensing device to the system?

4. If the cut-out point of a pressure control were set correctly, but the differential set too high, what would the symptoms be?

5. If a control were cutting in at the correct setting and the differential were set too high, what would the symptoms be?

LAB 14 Controlling the Head Pressure in a Water-Cooled Refrigeration System

Name ______________________ Date ____________ Grade ________

OBJECTIVES: Upon completion of this exercise, you will be able to set a water regulating valve for a water-cooled condenser and set the cut-out and cut-in point for a high-pressure control in the same system.

INTRODUCTION: You will control the head pressure in a water-cooled refrigeration (or air conditioning) system. This system will have a waste water condenser and a water regulating valve to control the head pressure. You will set the high-pressure control for protection against a water shut-off.

TEXT REFERENCES: Unit 10.

TOOLS AND MATERIALS: A gage manifold, valve wrench, goggles, gloves, and a water-cooled, waste water refrigeration system with a water regulating valve.

SAFETY PRECAUTIONS: You will be working with refrigerant, so care should be taken. Wear your goggles and gloves. NEVER USE A JUMPER TO TAKE A HIGH PRESSURE CONTROL OUT OF A SYSTEM AS EXCESSIVE HIGH PRESSURE MAY OCCUR.

PROCEDURES

1. See text for an example of a water regulating valve.
2. Fasten the high- and low-pressure gages on the system.
3. Open the service valves to allow pressure to the gages.
4. Purge a small amount of refrigerant from the center line to clean the gage lines.
5. Start the system.
6. After 5 minutes of running time, record the following:
   - Suction pressure __________ psig
   - Discharge pressure __________ psig
7. Turn the water regulating valve adjustment clockwise 2 turns, wait 5 minutes and record the following:
   - Suction pressure __________ psig
   - Discharge pressure __________ psig
8. Turn the water regulating valve adjustment counterclockwise 4 turns, wait 5 minutes and record the following:
   - Suction pressure __________ psig
   - Discharge pressure __________ psig
9. After completion of the above, slowly turn the adjustment stem of the water regulating valve in the direction that causes the head pressure to rise until the unit stops because of head pressure. CAUTION: DO NOT LET THE PRESSURE RISE ABOVE THE FOLLOWING:
   - R–12 147 psig or 115°F condensing temperature
   - R–22 243 psig or 115°F condensing temperature
   - R–502 263 psig or 115°F condensing temperature

10. If the compressor stops before the above settings, it is set correctly.

11. If the high-pressure control does not stop the compressor before these pressures, slowly turn the range adjustment on the high-pressure control to stop the compressor.

12. After completion of checking and setting the high-pressure control, reset the water pressure regulator to maintain the following pressures:

   - R–12 127 psig or 105°F condensing temperature
   - R–22 211 psig or 105°F condensing temperature
   - R–502 230 psig or 105°F condensing temperature

13. Remove the gage manifold.

MAINTENANCE OF WORK STATION AND TOOLS: Return all tools to their respective places. Place the system back in the manner in which you found it.

SUMMARY STATEMENT: Describe what would happen to the waterflow if the condenser tubes became dirty.

QUESTIONS

1. Why is a refrigerant cylinder not used to set a high-pressure control?

2. What is the purpose of the high-pressure control?

3. What is the purpose of the water regulating valve on a waste water system?

4. What senses the pressure of the water regulating valve?

5. Where is the typical water regulating valve connected to the system?

6. Where is the high-pressure control typically connected to the refrigeration system?

LAB 15 Identification of Electric Motors

Name ______________________________ Date ______________ Grade ________

OBJECTIVES: Upon completion of this exercise, you will be able to identify different types of electric motors, state their typical applications and sketch a diagram showing how they are wired into circuits.

INTRODUCTION: You will be provided an air conditioning unit and a refrigeration unit. You will remove enough panels on each system for you to identify the motors. You will draw a diagram showing how the motor is wired into the circuit.

TEXT REFERENCES: Unit 13.

TOOLS AND MATERIALS: A VOM (volt-ohm-milliammeter), straight blade and Phillips screwdrivers, 1/4" and 5/16" nut drivers and a flashlight.

SAFETY PRECAUTIONS: Turn the power off before removing any panels. Lock and tag the distribution box where you turned the power off and keep the single key in your possession.

PROCEDURES

Air Conditioning Unit

(Turn off the power. Lock and tag the box. Keep the key in your possession.)

1. Remove the panel to the evaporator section, exposing the indoor fan motor.
2. Using a good light, describe the following:
   - Number of wires entering the fan motor terminal box __________
   - Color of the wires __________
   - Is there a run capacitor? __________
   - Is there a start capacitor? __________
   - Motor voltage __________
   - Type of motor (capacitor-start or other type) __________
3. Replace the panels with the correct fasteners.
4. Remove the panel to the condenser section, exposing the outdoor fan motor.

5. Using a good light, describe the following:
   - Number of wires entering the fan motor terminal box __________
   - Color of the wires __________
   - Is there a run capacitor? __________
   - Is there a start capacitor? __________
   - Motor voltage __________
   - Type of motor (capacitor-start or other type) __________
6. Follow all wires going into the compressor and describe the following:
   - Number of wires entering the box __________
   - Is there a run capacitor? __________
   - Is there a start capacitor? __________
   - Is there a start assist (relay or a PTC device)? __________
   - Type of compressor motor (capacitor-start or other type) __________
7. Replace the panels with the correct fasteners.

MAINTENANCE OF WORK STANDARD AND TOOLS: Return all tools to their respective places.

SUMMARY STATEMENT: Describe the difference between a shaded-pole and a permanent split-capacitor fan motor. Explain which is the most efficient.

QUESTIONS

1. What is the reason for a start capacitor on a motor?

2. Must a motor with a start capacitor have a start relay? Why?

3. What is the purpose of a centrifugal switch in the end of an open motor?

4. Why can't you have a centrifugal switch in a hermetic compressor motor?

5. Name two methods used in hermetic compressor motors to take the start winding out of the circuit when the motor is up to speed.

6. How is a start capacitor constructed?

7. How is a run capacitor constructed?

8. How are capacitors rated?

9. Can a 230-V capacitor be used for a 440-V application?

10. Can a 440-V capacitor be used for a 230-V application?

11. Describe a PTC device.

12. How is a typical open electric motor cooled?

13. How is a typical hermetic compressor motor cooled?

LAB 16 Compressor Winding Identification

Name ______________________________ Date ______________ Grade ________

OBJECTIVES: Upon completion of this exercise, you will be able to identify the terminals to the start and run windings in a single-phase hermetic compressor.

INTRODUCTION: You will use VOM and learn the difference in the resistance of the start and run windings of a single-phase hermetic motor.

TEXT REFERENCES: Unit 13.

TOOLS AND MATERIALS: A VOM, straight blade and Phillips screwdrivers, 1/4" and 5/16" nut drivers.

SAFETY PRECAUTIONS: Turn the power off to any piece of equipment you are servicing. Lock the distribution box where you turned the power off and keep the single key in your possession.

PROCEDURES

1. You will be provided a single-phase compressor, either in a piece of equipment or from stock.
2. Turn the power off if the compressor is in a piece of equipment. Lock and tag the panel or box keeping the single key on your person.
3. Remove the compressor terminal cover.
4. If the compressor is wired into the circuit, turn the power on with supervision of your instructor, and check the voltage at the terminals: from terminal to terminal and terminal to ground.
5. With the power off, remove the terminal wiring one at a time and label the wires so that you can replace them correctly. NOTE: THE CROSSING OF THE WIRES CAN DO PERMANENT DAMAGE BEFORE YOU CAN SHUT THE COMPRESSOR OFF. DO NOT MAKE A MISTAKE.
6. Set the ohmmeter selector switch to R × 1 and 0 the meter.
7. Draw a diagram of the terminal layout, Figure 1.

   Record the following:

   1 to 2 __________Ohms

   1 to 3 __________Ohms

   2 to 3 __________Ohms

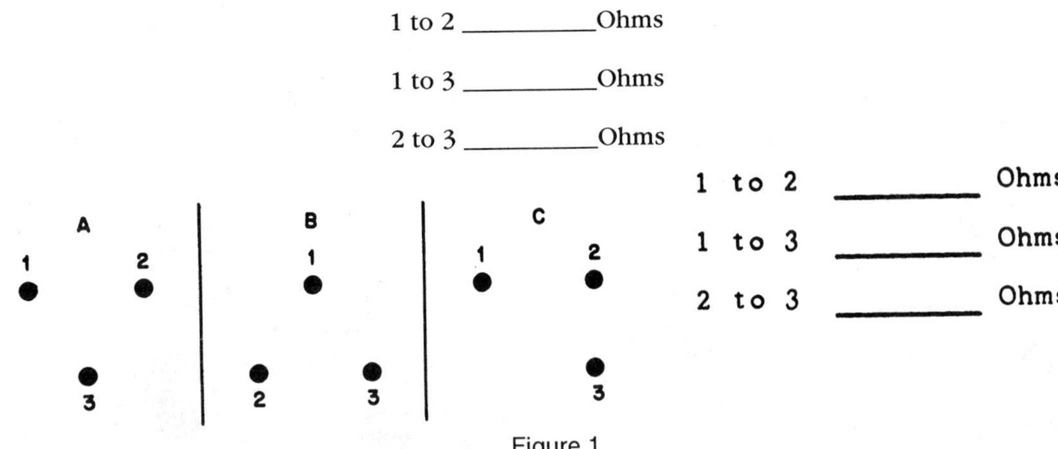

Figure 1

8. Based on your resistance readings, identify the start winding: terminal __________ to __________.
9. Based on your resistance readings, identify the run winding: terminal __________ to __________.

MAINTENANCE OF WORK STATION AND TOOLS: Return all equipment and tools to their respective places. Replace all panels with correct fasteners if appropriate.

SUMMARY STATEMENT: Describe the function of the start and run winding in a single-phase compressor.

QUESTIONS

1. Name the three terminals on a single-phase compressor.

2. Name 2 types of relays used to start a single-phase compressor motor.

3. Why does a hermetic compressor not use a centrifugal switch like an open motor?

4. Behind which terminal is the internal thermal overload located in a hermetic compressor?

5. What must be done to repair a compressor with a defective internal overload?

6. Which winding has the most resistance, the run or the start?

7. What cools the motor in a hermetic compressor?

8. Does oil ever touch the windings of a hermetic compressor?

9. Is there any difference in the materials in a hermetic compressor and an open motor?

10. What is the approximate loaded speed of a 3600-rpm compressor?

LAB 17 Motor Applications and Features

Name ______________________________ Date ______________ Grade ________

OBJECTIVES: Upon completion of this exercise, you will be able to fully describe a basic motor used for a fan in an air handler.

INTRODUCTION: You will remove a motor from an air handler, inspect it, describe the type of motor and describe its features, such as the speeds, insulation class, shaft size, rotation, mount characteristics, voltage, amperage, phase and bearing type.

TEXT REFERENCES: Unit 14.

TOOLS AND MATERIALS: A VOM, adjustable wrench, straight blade and Phillips screwdrivers, an Allen head wrench set, and 1/4" and 5/16" nut drivers.

SAFETY PRECAUTIONS: Turn the power off to any unit you are servicing. Lock and tag the power distribution panel where you turned the power off and keep the single key in your possession. Do not force any parts that you are disassembling or assembling, as they are made to work together and should not be bent or damaged in any way.

PROCEDURES

1. Make sure that the power is off to the air handler to which you have been assigned, that the power panel is locked and tagged, and you have the key. Remove the panels.
2. Carefully remove the fan section, fan and motor as an assembly.
3. Remove the motor from the mounting and fill in the following information:
   - Number of motor speeds __________
   - Insulation class of motor __________
   - Size of the motor shaft __________in.
   - Rotation looking at the shaft __________
   - Type of motor mount __________
   - Voltage rating __________V
   - Full-load amperage __________A
   - Locked-rotor amperage __________A
   - Phase of motor __________
   - Type of motor (PSC, shaded-pole) __________
   - Type of bearings (sleeve or ball) __________
   - Number of motor leads __________
4. Replace the motor in its mount.
5. Replace the fan section back into its unit. If any wires were removed, replace them as they were or as they should be.
6. Replace panels with the correct fasteners.

MAINTENANCE OF WORK STATION AND TOOLS: Return all tools to their respective places. Make sure that your work area is clean.

SUMMARY STATEMENTS: Describe how the motor was used in the exercise above. Explain if it is a belt- or direct-drive motor and state which is better for the application.

QUESTIONS

1. Why are sleeve bearings preferable for some fan applications?

2. What are the advantages of a ball bearing motor?

3. What is the name of the plug on the bottom of a bearing with a grease connection?

4. Do all bearings have to be lubricated each year?

5. What does resilient mount mean when applied to a motor?

6. Why must resilient-mount motors have a ground strap?

7. What are the upper and lower voltages at which a motor rated at 230 volts may be safely operated?
__________ lower, __________ upper

8. What are the upper and lower voltages at which a motor rated at 208 volts may be safely operated?
__________ lower, __________ upper

9. If a motor were to be operated below its rated voltage for long periods of time, what would happen?

10. Why do some motors have a different insulation class than others?

LAB 18 Identification of a Relay, Contactor, and Starter

Name ________________________________ Date ______________ Grade ________

OBJECTIVES: Upon completion of this exercise, you will be able to state the differences between a relay, a contactor, and a starter, and describe the individual parts of each.

INTRODUCTION: You will disassemble, inspect the parts, and reassemble a relay, a contactor, and a starter. You will compare the features of each.

TEXT REFERENCES: Unit 15.

TOOLS AND MATERIALS: Straight blade and Phillips screwdrivers, VOM, relay, contactor, and starter.

SAFETY PRECAUTIONS: Disassemble these devices carefully and reassemble them correctly.

PROCEDURES

1. Disassemble all parts of the relay. Lay the parts out in order on the bench. You may only be able to remove the cover.
2. Disassemble the contactor. Lay the parts out in order on the workbench. Remove the coil, the contacts (both movable and stationary), and the springs.
3. Disassemble the starter. Lay the parts out on in order on the bench. Remove the contacts, both the movable and the stationary, the coil and the overloads.
4. Assemble the relay back together.
5. Assemble the contactor. Check to make sure that all contacts are functioning, that they close when the armature is pushed up through the coil. Make sure that all screws holding the stationary contacts are tight.
6. Assemble the starter. Check to make sure that all contacts are functioning, that they close when the armature is pushed through the coil. Make sure that all screws holding the stationary contacts and the overloads are tight.

MAINTENANCE OF WORK STATION AND TOOLS: Return all tools and parts to their proper places. Make sure that your work area is left clean.

SUMMARY STATEMENTS:

A. State an application suitable for a relay where a contactor is not needed, and describe why.

B. State an application suitable for a contactor where a relay cannot be used, but a starter is not necessary. Tell why.

QUESTIONS

1. What is the difference between a relay and a contactor?

2. What is the difference between a contactor and a starter?

3. What should be done with a defective relay?

4. Are magnetic overloads fastened to a starter?

5. What effect would a loose connection have on a solder-pot-type overload?

6. What effect would a loose connection have on a bimetal overload?

7. What effect would a loose connection have on a magnetic-type overload?

8. Which of the following overloads is not affected by ambient temperature? (solder-pot, bimetal, or magnetic)

9. What is the difference between a relay and a potential relay?

10. What is the contact surface of a typical contactor made of?

LAB 19 Belt-Drive Motors

Name ______________________________ Date ______________ Grade ________

OBJECTIVES: Upon completion of this exercise, you will be able to disassemble and reassemble a belt-drive fan and motor and align the pulleys.

INTRODUCTION: You will remove a belt-drive fan section from an air handler, take the motor and pulley off, and remove the pulley from the fan shaft. You will then reassemble the fan section, align the pulleys, and set the belt tension.

TEXT REFERENCES: Unit 16.

TOOLS AND MATERIALS: A VOM, pulley puller, two adjustable wrenches, 3/8" – 1/2", 9/16" – 5/8" open or box end wrenches, oil can with light oil, straight blade and Phillips screwdrivers, straightedge, emery cloth (300 grit), soft face hammer, and 1/4" and 5/16" nut drivers.

SAFETY PRECAUTIONS: Turn the power off to the unit and lock and tag the panel before you start to disassemble it.

PROCEDURES

1. You will be assigned an air handler or furnace with a belt-drive fan motor. Turn the power off. Lock and tag the panel and keep the single key in your possession. Check the voltage with a meter to insure that the power is off.
2. Take off enough panels to remove the fan section.
3. Remove the fan section. Label any wiring that is disconnected.
4. Release the tension on the belt by loosening the adjustment.
5. Remove the belt and then the pulley from the motor shaft. DO NOT FORCE IT OR HAMMER THE SHAFT.
6. Remove the pulley from the fan shaft. DO NOT DISTORT THE SHAFT BY FORCING IT OFF.
7. Clean the fan shaft and motor shaft with fine emery cloth, about 300 grit.
8. Wipe, clean, and oil slightly.
9. Assemble the fan pulley to the fan shaft and the motor pulley to the motor shaft.
10. Place the motor back in its bracket and tighten it finger tight.
11. Align the pulleys as per the text.
12. Tighten the motor adjustment for tightening the belt, as per the text. Then tighten the motor onto its base.
13. Place the fan section back into the air handler and connect the motor wires.
14. Give the compartment a visual inspection and turn the fan motor over by hand. Make sure no wires or obstructions are in the way.

15. After your instructor has inspected and approved your work, shut the fan compartment door, turn the power on, and start the fan. Listen for any problems.

MAINTENANCE OF WORK STATION AND TOOLS: Wipe up any oil, clean the work station, and return all tools to their places.

SUMMARY STATEMENT: Describe your experience of removing the fan shaft pulley from its shaft.

QUESTIONS

1. Describe the effects of a belt that is too tight.

2. Why must pulleys be aligned accurately?

3. Why must a fan or motor shaft not be hammered?

4. State the advantages of a belt-drive motor over a direct-drive motor.

5. Do pulleys ever wear to the point of needing replacement?

6. How do you adjust a pulley for a faster-driven pulley speed?

7. Are all belt and pulley groves alike?

8. What is a matched set of belts?

9. What are fan shafts made of?

10. Do all fan motors turn at the same speed?

LAB 20 Examination of a Hermetic Compressor Electric Motor

Name ______________________________ Date ______________ Grade _______

OBJECTIVES: Upon completion of this exercise, you will be able to evaluate a motor in a hermetic compressor to determine if it is electrically sound and safe to start.

INTRODUCTION: You will use an ohmmeter to check a hermetic compressor from terminal to terminal and from terminal to ground to determine if the motor is safe to operate.

TEXT REFERENCES: Unit 16.

TOOLS AND MATERIALS: A VOM, straight blade and Phillips screwdrivers, and 1/4" and 5/16" nut drivers.

Safety Precautions: Make sure that the power is turned off before beginning this exercise. Lock and tag the panel where the power is turned off and keep the single key on your person.

PROCEDURES

1. You will be assigned a hermetic compressor to use for this exercise. If the compressor is part of a system, turn the power off, and lock and tag the panel.
2. Label the terminal wires and remove them from the terminals.
3. Turn the range selector switch on the VOM to the R × 1 scale. Fasten the two leads together and 0 the meter.
4. Determine which terminals are Common, Run, and Start, either by using the resistance method or by the identification in the terminal box.
5. Fasten one lead on the VOM to the Common terminal.
6. Touch the other lead to the Run terminal and record the reading here. __________Ohms
7. Now, touch the lead to the Start terminal and record the reading here. __________Ohms
8. If you have a book that lists this compressor and gives the winding resistance, compare the ones you took with the book.
9. Turn the ohmmeter selector switch to the R × 1000 and 0 the meter again.
10. Fasten one lead to a good frame ground on the compressor; the copper process tube or discharge line will do. Touch the other lead to another part of the compressor frame. The meter should read 0 resistance.
11. With one lead on the ground, touch the other lead to each of the three terminals. The reading should be the same for each terminal. Record here. __________Ohms
12. If the compressor has a winding thermostat with leads wired to external terminals, with one meter lead attached to the ground, touch the other meter lead to each of them and record the reading here. __________Ohms The winding thermostat should also be checked to the motor terminals by placing one lead on the winding thermostat and the other on each of the motor terminals. Record the readings here. __________Ohms, __________Ohms, __________Ohms

MAINTENANCE OF WORK STATION AND TOOLS: Replace all panels that may have been removed and put all tools in their respective places.

SUMMARY STATEMENT: Describe a circuit to ground in a motor.

QUESTIONS

1. What would be the results of starting a motor that is grounded?

2. How would a motor that has an open run winding react at starting time?

3. How would a motor that has an open start winding react while starting?

4. How should the resistance reading compare in each winding of a three-phase motor?

5. If a motor's windings all had the correct resistance reading, the correct power was supplied to the windings, and the compressor would not start, what would be the probable cause?

6. Describe a shunted winding.

7. What voltage does a potential relay operate from?

8. Describe how a current relay operates when removing the start winding from the circuit.

LAB 21 Checking a Capacitor with an Ohmmeter

Name ______________________________ Date ______________ Grade ______

OBJECTIVES: Upon completion of this exercise, you will be able to field check a start or run capacitor using an ohmmeter.

INTRODUCTION: You will use the ohmmeter feature of a VOM to check a capacitor. This will not give the actual capacitance of a capacitor, but will show a short, open, or grounded capacitor. The procedures in this exercise can be demonstrated better using an ohmmeter with a needle scale.

TEXT REFERENCES: Unit 16.

TOOLS AND MATERIALS: A straight blade screwdriver; VOM; insulated needle nose pliers; 20,000-ohm, 5-watt resistor; and a start and run capacitor.

SAFETY PRECAUTIONS: Always short a capacitor from terminal to terminal before applying an ohmmeter or the ohmmeter may be damaged. The capacitor should be shorted with a 20,000-ohm, 5-watt resistor. Use insulated needle nose pliers.

PROCEDURES

1. You will be given a start and a run capacitor either from stock or from a unit. If you are going to remove it from a unit, TURN THE POWER OFF. Lock and tag the panel while the power is off. Keep the only key in your possession.
2. Short the capacitor from terminal to terminal. See text for an explanation of how this procedure is done.
3. Set the range selector switch on the meter to ohms and the R × 100 scale.
4. Using the run capacitor, touch one lead on each terminal of the run capacitor while watching the meter. If you get no response, reverse the leads. The needle should rise and fall. It will finally fall back to ∞ if the capacitor is good. If the needle only falls back to a point, the capacitor has an internal short of the value showing on the meter.
5. Now change the ohmmeter range selector switch to R × 1000.
6. Place one lead on a capacitor terminal and touch the other lead to the meter can. Any needle movement at all indicates that the capacitor is shorted to the can. It is defective.
7. Did the run capacitor show "good" in this test?
8. Using the start capacitor, short between the terminals. NOTE: If the capacitor has a resistor between the terminals, it is there for the purpose of bleeding the capacitor charge. There should be no stored charge if the resistor is good.
9. Turn the ohmmeter range selector switch to the R × 100 scale.
10. Fasten one meter lead to one terminal and then touch the other meter lead to the other terminal. The meter needle should rise and fall back. IT WILL FALL BACK MUCH SLOWER THAN WITH THE RUN CAPACITOR BECAUSE OF THE GREATER CAPACITANCE. NOTE: The needle will only fall back to the value of the bleed resistor. If you need to know if the needle will fall all the way back to infinity, the bleed resistor must be disconnected.

11. You will not perform a ground check on a start capacitor because the case is plastic and a non-conductor.

12. Did the start capacitor show "good" in this test?

MAINTENANCE OF WORK STATION AND TOOLS: Return all parts and tools back to their respective places. Make sure that your work area is left clean.

SUMMARY STATEMENTS: Describe how a capacitor functions in a starting circuit and in a running circuit.

QUESTIONS

1. What is a start capacitor container made of?

2. What is a run capacitor container made of?

3. What is the fill material in a start capacitor?

4. What is the fill material in a run capacitor?

5. What is the purpose of a bleed resistor in a start capacitor?

6. What would the symptoms be of an open start capacitor if the compressor were started?

7. What would the symptoms be of a shorted run capacitor if the compressor were started?

8. What is the purpose of the run capacitor in a motor circuit?

LAB 22 Comfort Conditions

Name ______________________________ Date ______________ Grade ________

OBJECTIVES: Upon completion of this exercise, you will be able to take wet-bulb and dry-bulb temperature readings, determine relative humidity from the psychrometric chart, and use this information to determine the level of comfort from the ASHRAE generalized comfort chart.

INTRODUCTION: With a sling psychrometer you will take several dry- and wet-bulb temperature readings in a conditioned space. By referring to a psychrometric and generalized comfort chart you will determine if the conditioned space is within accepted comfort limits.

TEXT REFERENCES: Unit 17.

TOOLS AND MATERIALS: A sling psychrometer, and the comfort and psychrometric charts. See text.

SAFETY PRECAUTIONS: The sling psychrometer is a fragile instrument. Be careful not to break it. Keep away from other objects when using it.

PROCEDURES

1. Review procedures for using the sling psychrometer.

2. Record the following for 10 wet-bulb/dry-bulb readings.

	Dry Bulb	Wet Bulb	Relative Humidity (from psychrometric chart)	Within Generalized Comfort Zone (yes or no)
1.	________	________	________	________
2.	________	________	________	________
3.	________	________	________	________
4.	________	________	________	________
5.	________	________	________	________
6.	________	________	________	________
7.	________	________	________	________
8.	________	________	________	________
9.	________	________	________	________
10.	________	________	________	________

MAINTENANCE OF WORK STATION AND TOOLS: Return the sling psychrometer to its storage space.

SUMMARY STATEMENTS: Describe the general comfort level of the conditioned space you evaluated. What air conditioning adjustments can be made to make the space more comfortable?

QUESTIONS

1. What causes the thermometer with the wet sock to read lower than the thermometer with the dry bulb?

2. Why must the thermometer be swung in order to obtain the reading?

3. Name the 5 things that can be done to condition air.

4. Name the 4 comfort factors.

5. How does air movement improve comfort?

6. What is the difference in the summer and winter comfort chart readings?

7. Name three different kinds of air filters.

8. What does it mean when it is said that air is 70% RH?

9. Name 2 ways that humidity can be added to air.

10. Name 3 ways that the human body gives up heat.

LAB 23 Window Air Conditioner Familiarization

Name ______________________________ Date ______________ Grade __________

OBJECTIVES: Upon completion of this exercise, you will be able to identify the various parts of a window air conditioner.

INTRODUCTION: Your instructor will assign you a window unit; you will remove it from its case, identify, and describe the major components.

TEXT REFERENCES: Unit 18.

TOOLS AND MATERIALS: Straight blade and Phillips screwdrivers, 1/4" and 5/16" nut drivers, flashlight, and a window or room air conditioner.

SAFETY PRECAUTIONS: Make sure the unit is unplugged before starting the exercise. Be careful while working around the coil fans as they are sharp.

PROCEDURES

1. If the unit is installed in a window in a slide out case, remove it from its case and set it on a level bench. If the unit is in a case fastened with screws, set the unit on a bench and remove enough screws to remove the case.

2. When you have the unit where all parts are visible, fill in the following information.

   INDOOR SECTION
   Type of fan blade __________ What is the evaporator tubing made of? __________ What are the fins made of? __________ Where does the condensate drain to from the evaporator? __________

   OUTDOOR SECTION
   Type of fan blade __________ What is the condenser tubing made of? __________ What are the fins made of? __________ Does the condenser fan blade have a slinger ring for evaporating condensate? __________ Does the condenser coil appear to be clean for a good heat exchange? __________

   FAN MOTOR INFORMATION
   Size of fan shaft __________ Looking at the lead end of the motor, what is the rotation, clockwise or counterclockwise? __________ Fan motor voltage __________ V Fan motor current __________ A Type of fan motor, PSC or shaded pole __________ If PSC, what is the MFD rating of the capacitor? __________ MFD If available on the nameplate of the motor, what are its speeds? __________

   UNIT NAME PLATE
   Manufacturer __________ Serial number __________ Model number __________ Unit voltage __________ Unit full load current __________ A Test pressure __________ psi.

   GENERAL UNIT INFORMATION
   What ampere rating does the power cord have? __________ A Was this unit intended for a casement window or a double hung window? __________ What type of air filter is furnished with the unit? (fiberglass or foam rubber) __________.

3. Assemble the unit.

MAINTENANCE OF WORK STATION AND TOOLS: Replace all tools in their proper location. Place the window or room unit in a storage area if instructed. Leave your work area clean.

SUMMARY STATEMENT: Describe how condensate is evaporated in the condenser portion of the window unit you worked with.

QUESTIONS

1. How many fan motors are normally included in a window cooling unit?

2. What are the two types of designs of window units relative to covers and access for service?

3. What is the most common refrigerant used in window units?

4. What is the most common metering device used in window units?

5. What is the purpose of the heat exchange between the metering device and the suction line?

6. Why must window cooling units operate above freezing?

7. At what temperature do window cooling unit evaporators typically boil the refrigerant?

8. What is the evaporator tubing in these units usually made of?

9. What are the evaporator fins made of?

10. How does the unit dehumidify the conditioned space?

LAB 24 Checking the Charge without Gages

Name ______________________________ Date ______________ Grade __________

OBJECTIVES: Upon completion of this exercise, you will be able to evaluate the refrigerant charge for a window air conditioner without disturbing the refrigerant circuit with gage connections.

INTRODUCTION: You will operate a window unit and use temperature drop across the evaporator and temperature rise across the condenser to evaluate the refrigerant charge without gages.

TEXT REFERENCES: Unit 18.

TOOLS AND MATERIALS: Electronic thermometer (or 2 glass thermometers), straight blade and Phillips screwdrivers, 1/4" and 5/16" nut drivers, and window cooling unit.

SAFETY PRECAUTIONS: Make sure that when the unit panels are removed for the purpose of observing the suction line that the condenser and evaporator have full airflow over them. Insure that all leads, clothing, and your hands are kept away from fan blades.

PROCEDURES

1. Remove the unit from its case and place it on a bench with a correct power supply nearby.
2. Make sure that the correct airflow passes across each coil before starting the compressor. Cardboard may be placed over any place where a panel has been removed to direct the airflow across the coil.
3. Place a thermometer at the inlet and outlet of the evaporator and start the compressor. DON'T LET THE THERMOMETER LEAD GET IN THE FAN.
4. Cover about 2/3 of the condenser surface such as in Figure 1. This will cause the head pressure to rise. DO NOT COVER THE ENTIRE CONDENSER. When the head pressure rises, the condenser then operates with the correct charge, pushing the correct refrigerant charge to the evaporator. The suction line leading to the compressor should then sweat.

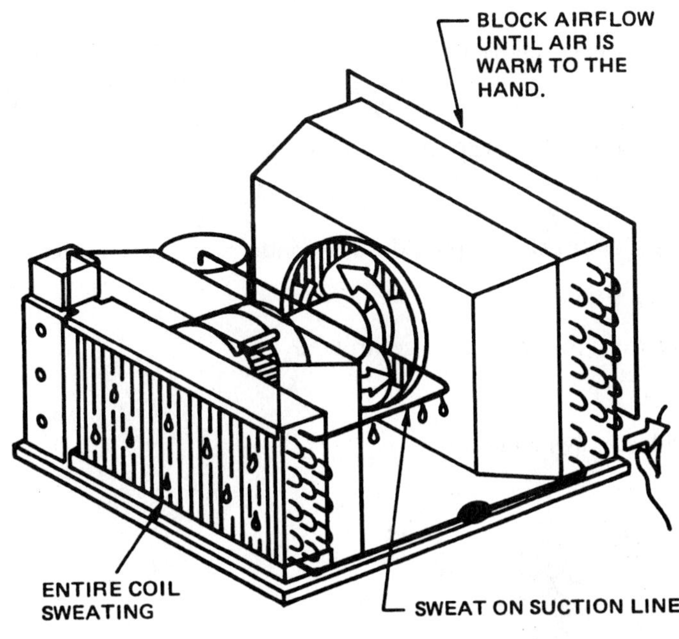

Figure 1

5. Let the unit run for about 15 minutes, then go to step 6.
6. How much of the evaporator is sweating, 1/4, 1/2, 3/4, or all of it? How much of the suction line to the compressor is sweating? __________
7. Record the air temperature entering the evaporator __________°F.
Record the air temperature leaving the evaporator __________°F.
Record the temperature difference __________°F.
8. Now record the temperature difference across the condenser __________°F. It will not be significant, but worth noting.
9. Shut the unit off and unplug. Replace all panels and place the unit in its permanent location.

MAINTENANCE OF WORK STATION AND TOOLS: Clean up the work station and return all tools to their proper places.

SUMMARY STATEMENTS: Evaluate the unit you worked on and tell how the above test indicates whether the charge is correct.

QUESTIONS

1. What is the primary maintenance needed on room cooling units?
2. How do some manufacturers provide evaporator freeze protection?
3. Why should gages be installed only when absolutely necessary?
4. What two types of compressors may be used in room units?
5. What are the four major components in the room unit refrigeration cycle?
6. What are two purposes of the condenser?
7. How is the compressor heat used to evaporate condensate?
8. Describe how the condenser fan slinger provides for condensate evaporation.
9. How can the exhaust or fresh air options be controlled?
10. What are options the selector switch controls?

LAB 25 Charging a Window Unit Using the Sweat Line Method

Name ______________________________ Date ______________ Grade __________

OBJECTIVES: Upon completion of this exercise, you will be able to add a partial charge to a window unit using the sweat line method.

INTRODUCTION: You will add refrigerant to a window unit through the suction line service port until a correct pattern of sweat has formed on the suction line to the compressor.

TEXT REFERENCES: Unit 18.

TOOLS AND MATERIALS: Straight blade and Phillips screwdrivers, 1/4" and 5/16" nut drivers, gage manifold, cylinder of refrigerant (R-22 for most units), gloves, goggles, and a window unit with low and high side pressure service ports.

SAFETY PRECAUTIONS: Gloves and goggles should be worn while making gage connections and transferring refrigerant. A high side gage should be used while adding refrigerant.

PROCEDURES

1. Unplug the unit and remove it from its case. Set the unit on a bench with a power supply for the unit.
2. Put on goggles and gloves. Fasten the gage lines to the high and low side connections. Purge the gage lines through the center line and connect it to the refrigerant cylinder.
3. Make sure the airflow across the evaporator and condenser is correct. Start the unit and observe the pressures. Record the following after 15 minutes: Suction pressure __________ psig. Discharge pressure __________ psig. Where is the last point that the suction line is sweating?
4. Restrict the airflow across the condenser as in Lab 24 until about 2/3 of the condenser is blocked. Wait 15 minutes and record the following: Suction pressure __________ psig Discharge pressure __________ psig. Where is the last point the suction line is sweating?
5. If the unit has the correct charge, allow some refrigerant to escape the system into the refrigerant cylinder through the discharge line. Let refrigerant out until the suction pressure is about 30 psig for R-22.
6. Record the following after 15 minutes of running time at about 30 psig of suction pressure. Suction pressure __________ psig Discharge pressure __________ psig Where is the last point that the suction line is sweating?
7. Start adding vapor refrigerant in short bursts through the suction line. Add a little and wait for about 5 minutes. Keep adding in short bursts until the sweat line moves to the compressor, waiting about 5 minutes after each addition of refrigerant. When the correct charge is reached, record the following: Suction pressure __________ psig Discharge pressure __________ psig. See Figure 1.
8. Unplug the unit and remove the gages. Return the unit to its permanent location. Be sure to seal the unit gage ports.

MAINTENANCE OF WORK STATION AND TOOLS: Return all tools to their proper location and make sure the work station is clean.

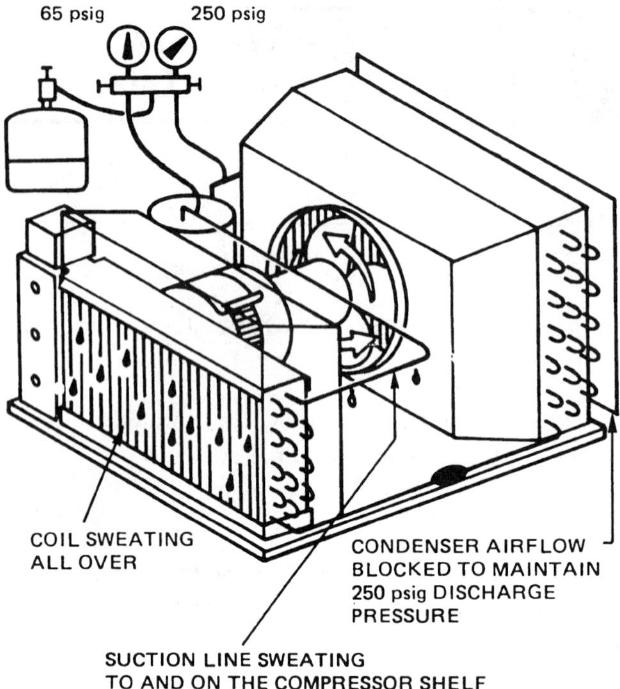

Figure 1 Charging a unit using the sweat line method

SUMMARY STATEMENT: Explain why the condenser airflow must be blocked in order to correctly add a partial charge to a unit.

QUESTIONS

1. Does the fan motor normally need periodic lubrication?

2. What will happen to the suction pressure if the indoor coil becomes dirty?

3. If the low-side pressure becomes too low, what will happen to the condensate on the evaporator coil?

4. Why should gages be installed only when absolutely needed?

5. What are line tap valves?

6. In this exercise, why is the airflow through the condenser blocked?

7. If only part of the evaporator gets cold in this test, what could be the problem?

8. If the charge is correct, but the evaporator coil is only cool, what could be the problem?

LAB 26 Charging a Unit Using Scales or a Charging Cylinder

Name ______________________________ Date ______________ Grade ________

OBJECTIVES: Upon completion of this exercise, you will be able to add an accurate charge to a window unit using scales or a charging cylinder.

INTRODUCTION: You will evacuate a unit to a low vacuum and add a measured charge to the unit using scales or a charging cylinder.

TEXT REFERENCES: Unit 18.

TOOLS AND MATERIALS: Straight blade and Phillips screwdrivers, 1/4" and 5/16" nut drivers, gage manifold, cylinder of correct refrigerant for the system, vacuum pump, goggles, gloves, and a room cooling unit.

SAFETY PRECAUTIONS: Goggles and gloves should be worn while making gage connections and transferring refrigerant.

PROCEDURES

1. Remove the unit to a bench with power. Put on goggles and gloves. Fasten the gage lines to the gage ports and purge the gage lines. NOTE: IF THE UNIT HAS ONLY A SUCTION PORT YOU CAN STILL DO THIS EXERCISE.
2. If the unit has a charge of refrigerant, your instructor may advise you to recover the refrigerant from the system using an approved recovery system.
3. When all of the refrigerant is removed from the system, unplug the unit and connect the vacuum pump. Check the oil in the vacuum pump. If it is not at the correct level, ask your instructor for directions. Start the pump to evacuate the system.
4. While the vacuum pump is running, get set up to charge the system either with scales or a charging cylinder. Determine the correct charge, make necessary calculations, and be ready to add the refrigerant as soon as the correct vacuum has been obtained.
5. NOTE: IF YOU ARE PULLING A VACUUM FROM BOTH THE LOW AND HIGH PRESSURE SIDES OF THE SYSTEM, YOU SHOULD SHUT THE HIGH PRESSURE SIDE OF THE SYSTEM OFF AND REMOVE THE GAGE LINE WHEN A DEEP VACUUM HAS BEEN REACHED. IF THE GAGE PORT IS A SCHRADER VALVE, ALLOW THE PRESSURE IN THE SYSTEM TO RISE TO ABOUT 10 PSIG WHILE ADDING REFRIGERANT AND THEN REMOVE THE GAGE LINE. This procedure is to keep the measured refrigerant from moving into the high pressure line and condensing, and affecting the charge.
6. When refrigerant is allowed to enter the system, the system pressure will soon equalize with the cylinder pressure and no more refrigerant will flow. You may start the unit and the rest of the charge may be charged into the system. NOTE: IF YOU ONLY HAVE A LOW PRESSURE CONNECTION, BE CAREFUL THAT YOU ONLY CHARGE THE CORRECT AMOUNT INTO THE SYSTEM.
7. When the correct charge is in the system, shut the unit off and disconnect the low side gage line from the unit. Place caps on all gage connections and replace the panels.
8. Return the unit to its permanent location.

MAINTENANCE OF WORK STATION AND TOOLS: Return all tools to their proper location.

SUMMARY STATEMENTS: Describe what you found to be the most difficult step in the above charging sequence.

QUESTIONS

1. Why are larger ports better when evacuating a system?

2. What is the problem if when charging a system the head pressure rises and the suction pressure does not?

3. If there is a problem with the capillary tube, what types of metering devices may be installed in a room cooling unit?

4. What must the compressor motor have for starting devices if one of the above metering devices is installed?

5. Why must the motor have the above starting devices?

6. Why is it important to check for a leak if the refrigerant is found to be low?

7. Why should an accurate dial or electronic scale be used when weighing in the refrigerant?

LAB 27 Evaporator Functions and Superheat

Name ______________________________ Date ______________ Grade ________

OBJECTIVES: Upon completion of this exercise, you will be able to take pressure and temperature readings and determine superheat of the refrigerant in the evaporator operating under normal and loaded conditions.

INTRODUCTION: You will be working with an evaporator in an operating system. Any typical evaporator in a refrigeration system operating in its normal environment can be used as long as there are gage ports for both high- and low-side gages. You will be taking pressure and temperature readings.

TEXT REFERENCES: Unit 19.

TOOLS AND MATERIALS: A gage manifold, straight blade screwdriver, 1/4" and 5/16" nut drivers, light gloves, goggles, and thermometer that can be attached to the suction line, a small amount of insulation for thc thermometer, and an evaporator in an operating system.

SAFETY PRECAUTIONS: You will be working with refrigerant under pressure. Care should be taken not to allow the refrigerant to get in your eyes. Wear goggles at all times when working with refrigerants. Liquid refrigerant will freeze the skin. If a leak develops and liquid refrigerant escapes, DO NOT TRY TO STOP IT WITH YOUR HANDS. Light gloves are recommended when handling all refrigerants. Your instructor must give you instruction in using the gage manifold before beginning this exercise. Be careful of all moving parts and especially the evaporator fan.

PROCEDURES

1. Put on your goggles and gloves. Fasten the suction gage line to the suction gage port. The high-side line may be attached if desired; however, it is not necessary.

2. Fasten the thermometer to the suction line securely (electrical tape works well) and insulate the lead for about 4 inches in each direction, Figure 1.

3. Start the unit, allow it to run for at least 15 minutes, and then record the following.

   Suction pressure: ________ psig.

   Determine boiling temperature from the PT chart: ________°F

   Suction line temperature: ________°F

   Determine evaporator superheat:
   Suction line temperature – boiling temperature = ________°F superheat

4. Now, increase the load on the evaporator. If it is a reach-in box, open the door and place some room temperature objects in the box. If it is an air conditioner, partially open the fan compartment door to increase the airflow across the evaporator. See Figure 2. After 15 minutes, record the following:

   Suction pressure: ________ psig

   Boiling temperature from the PT chart: ________°F

   Suction line temperature: ________°F

   Determine evaporator superheat:
   Suction line temperature – boiling temperature = ________°F superheat

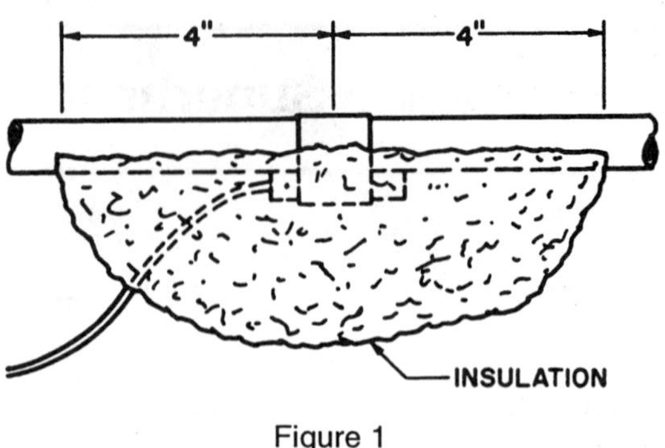

Figure 1

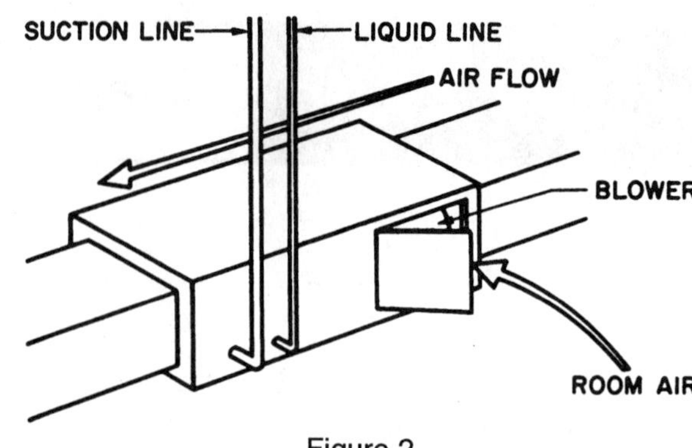

Figure 2

5. Remove the gages and thermometer and return the unit to normal running condition with all panels in place.

MAINTENANCE OF WORK STATION AND TOOLS: Wipe any oil off the gages and replace all plugs and caps. Return all equipment and tools to their respective places.

SUMMARY STATEMENT: Describe what happened when a load was placed on the evaporator.

QUESTIONS

1. What is the typical superheat for an evaporator operating under normal conditions? _________°F

2. Why is it important to have superheat at the end of the evaporator?

3. What is the job of the evaporator in a refrigeration system?

4. The refrigerant entering the evaporator is approximately _________% liquid and _________% vapor.

5. The refrigerant leaving the evaporator is _________% vapor.

6. Where does the refrigerant go when it leaves the evaporator?

7. The refrigerant in the evaporator must be (colder or warmer) than the air passing over it for heat to transfer into the coil.

LAB 28 Evaporator Performance

Name ______________________________ Date ______________ Grade ________

OBJECTIVES: Upon completion of this exercise, you will be able to evaluate the performance of a direct expansion evaporator.

INTRODUCTION: You will use superheat and coil-to-air temperature difference to evaluate a direct expansion evaporator.

TEXT REFERENCES: Unit 19.

TOOLS AND MATERIALS: An electronic thermometer, screwdriver, electrician's tape, some insulation, gage manifold, goggles, and refrigeration system with a thermostatic expansion valve and service valves.

SAFETY PRECAUTIONS: You will be fastening gages to the refrigeration system. Wear goggles and gloves. You must have had instruction in using the gage manifold before starting this exercise.

PROCEDURES

1. Put on your goggles and gloves. Fasten the high- and low-side gage lines to the service valves. Open the service valves by turning them off the back seat. Turn them clockwise.
2. Purge the air from the gage lines by loosening the fittings slightly at the manifold.
3. Fasten a thermometer lead to the suction line of the evaporator where the line leaves the coil. This will be in the same area as the TXV (thermostatic expansion valve) sensor. Be sure the temperature lead is insulated.
4. Fasten another lead in the return air of the evaporator.
5. Start the system and allow it to pull down to within 10°F of its intended use (low, medium, or high temperature) and record the following before the thermostat shuts the compressor off:
   - Suction pressure ________ psig
   - Temperature at the suction line ________°F
   - Refrigerant boiling temperature (converted from suction pressure) ________°F
   - Superheat (suction line temperature minus boiling temperature) ________°F
   - Return air temperature ________°F
   - TD (temperature difference between return air and refrigerant boiling temperature) ________°F
   - Discharge pressure ________ psig
   - Condensing temperature (converted from discharge pressure) ________°F
6. Back seat the service valves and remove the gage lines.
7. Remove the temperature leads and replace them in their case.

MAINTENANCE OF WORK STATION AND TOOLS. Replace all panels and return all tools to their respective places. Make sure the refrigerated box is returned to its correct condition.

SUMMARY STATEMENT: Describe how well the evaporator is performing in the above test.

QUESTIONS

1. What is a typical superheat for a typical direct expansion evaporator performing in the correct temperature range?

2. What is the difference between a direct expansion and a flooded evaporator?

3. What device is used to evenly feed refrigerant to an evaporator with more than one circuit?

4. What would the symptoms of a flooded evaporator be on a direct expansion system?

5. What would the symptoms of a dirty coil be?

6. Why is it necessary to insulate the bulb on the thermometer when taking the superheat of an evaporator?

7. What is the problem when the superheat reading is too low?

8. What is the problem when the superheat reading is too high?

9. What does the term "starved evaporator" mean?

10. When the superheat is too low, what can happen to the compressor?

LAB 29 Checking an Evaporator Under Excessive Load

Name ______________________ Date ____________ Grade ________

OBJECTIVES: Upon completion of this exercise, you will be able to check an evaporator while it is under excessive load and observe the changing conditions while the load is being reduced.

INTRODUCTION: You will be assigned a packaged commercial refrigeration system with a thermostatic expansion valve that has been turned off long enough to be warm inside. You will then turn it on with an ammeter, gages and thermometer attached, and record the conditions while the temperature is being lowered in the box.

TEXT REFERENCES: Unit 19.

TOOLS AND MATERIALS: A gage manifold, four-lead thermometer, ammeter, goggles, gloves, straight blade and Phillips screwdrivers, 1/4" and 5/16" nut drivers, and a commercial refrigeration system with thermostatic expansion valve.

SAFETY PRECAUTIONS: Wear goggles and gloves while working with pressurized refrigerant.

PROCEDURES

1. Put on your goggles and gloves. Connect the high- and low-side gage lines.
2. If the system has service valves, turn the valve stems off the back seat far enough for the gages to read. If the system has Schrader valves, the gages will read when connected.
3. Fasten the four temperature leads in the following locations:
   - Suction line leaving the evaporator (before a heat exchanger, if any). See the text for an example of evaporator pressures and temperature.
   - Inlet air to the evaporator
   - Air entering the condenser
4. Fasten the ammeter around the common wire on the compressor.
5. Record the following information two times (TEST 1 ten minutes after start up and TEST 2 when the box has reduced the temperature to near the shut-off temperature).
   - Type of refrigerant R-_____

	TEST 1	TEST 2
• Suction pressure	________	________psig
• Suction line temperature	________	________°F
• Evaporator boiling temperature	________	________°F
• Superheat	________	________°F
• Air temperature entering evaporator	________	________°F
• Evaporator TD (air entering the evaporator – refrigerant boiling temperature)	________	________°F
• Discharge pressure	________	________psig
• Air temperature entering condenser	________	________°F
• Condensing temperature	________	________°F
• Condenser TD (condensing temperature – entering air temperature)	________	________°F
• Compressor full-load amperage	________	________A

6. Remove all instruments and gages.

7. Replace all panels with the correct fasteners.

8. Return the system to the condition stated by the instructor.

MAINTENANCE OF WORK STATION AND TOOLS. Return all tools to their respective work stations.

SUMMARY STATEMENT: Describe how the amperage, suction, and discharge pressure responded as the temperature in the refrigerated box was lowered.

QUESTIONS

1. What is the function of the evaporator in a refrigeration system?

2. How does boiling refrigerant absorb heat when it is cold?

3. What should the typical superheat be at the end of the evaporator before the heat exchanger? _________°F

4. What is the purpose of the fins on an evaporator?

5. Where does the condensate come from that forms on the evaporator?

6. What is done with the condensate mentioned in question 5?

7. What is a dry-type evaporator?

8. What is the ratio of liquid to vapor refrigerant at the beginning of the evaporator? _________% liquid and _________% vapor

LAB 30 Compressor Operating Temperatures

Name ______________________ Date ____________ Grade ________

OBJECTIVES: Upon completion of this exercise, you will be able to evaluate a suction gas-cooled compressor for the correct operating temperatures while it is operating.

INTRODUCTION: You will use gages and a thermometer to become familiar with how a suction gas-cooled compressor should feel while operating, and where it should be hot, warm, and cool.

TEXT REFERENCES: Unit 20.

TOOLS AND MATERIALS: A gage manifold, straight blade and Phillips head screwdrivers, 1/4" and 5/16" nut drivers, goggles, gloves, thermometer, and a refrigeration system with a gas-cooled compressor.

SAFETY PRECAUTIONS: Wear goggles and gloves while working with refrigerant under pressure.

PROCEDURES

1. The refrigeration system should have been off for some period of time. It should have a welded hermetic compressor and a suction-cooled motor.
2. Put on your goggles and gloves. Connect the high- and low-side gages and obtain a gage reading. There must either be service valves or Schrader valve ports.
3. Fasten one temperature sensor to the suction line entering the compressor, Figure 1.

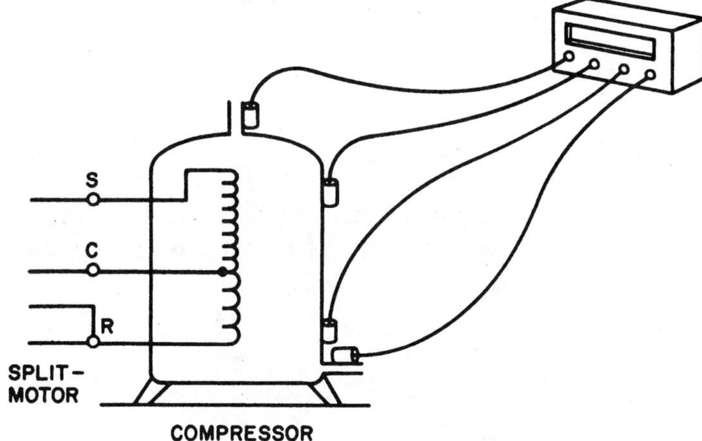

Figure 1 The correct location for the temperature sensors. The sensors may be fastened with electrical tape and should have some insulation.

4. Fasten a second temperature lead to the compressor shell about 4" from where the suction line enters the shell.
5. Fasten a third temperature sensor to the bottom of the compressor shell where the oil would be.
6. Fasten the fourth temperature sensor to the discharge line leaving the compressor.

7. Record the following data before starting the system:

- Type of refrigerant in the system __________
- Suction pressure __________psig
- Refrigerant boiling temperature __________°F
- Discharge pressure __________psig
- Refrigerant condensing temperature __________°F
- Suction line temperature __________°F
- Upper shell temperature __________°F
- Lower shell temperature __________°F
- Discharge line temperature __________°F

8. Start the system and let it run for 30 minutes and record the following information:

- Suction pressure __________psig
- Refrigerant boiling temperature __________°F
- Discharge pressure __________psig
- Refrigerant condensing temperature __________°F
- Suction line temperature __________°F
- Upper shell temperature __________°F
- Lower shell temperature __________°F
- Discharge line temperature __________°F

9. Block the condenser airflow until the refrigerant is condensing at 125°F. This would be one of the following:

- R-12 169 psig
- R-22 278 psig
- R-502 301 psig

10. Let the system run for 30 minutes at this condition. NOTE: YOU MAY HAVE TO TURN THE THERMOSTAT CONTROLLING THE UNIT DOWN TO KEEP IT FROM SHUTTING OFF DURING THIS RUNNING TIME. Record the following information:

- Suction pressure __________psig
- Refrigerant boiling temperature __________°F
- Discharge pressure __________psig
- Refrigerant condensing temperature __________°F
- Suction line temperature __________°F
- Upper shell temperature __________°F
- Lower shell temperature __________°F
- Discharge line temperature __________°F

11. Shut the system off for 15 minutes and record the following conditions again:
   - Suction pressure __________psig
   - Refrigerant boiling temperature __________°F
   - Discharge pressure __________psig
   - Refrigerant condensing temperature __________°F
   - Suction line temperature __________°F
   - Upper shell temperature __________°F
   - Lower shell temperature __________°F
   - Discharge line temperature __________°F
12. Reset the thermostat to the correct setting.
13. Remove the gages and thermometer sensors.

MAINTENANCE OF WORK STATION AND TOOLS: Return all tools to their respective places. Replace all panels. Leave the system in the manner in which you found it unless instructed otherwise.

SUMMARY STATEMENTS: Describe how the different temperatures of the compressor responded to a rise in head pressure and why.

Suction line —

Upper shell —

Lower shell —

Discharge line —

QUESTIONS

1. What would happen to the discharge line temperature with an increase in superheat at the suction line?

2. What would happen to the discharge line temperature with a decrease in suction line temperature?

3. What would the result of an internal discharge line leak be on a compressor's temperature?

4. What is the purpose of a crankcase heater on a compressor?

5. How is a typical hermetic compressor mounted inside the shell?

6. Is the compressor shell thought of as being on the high- or low-pressure side of the refrigeration system?

7. Is the suction or discharge line on the compressor typically larger?

8. What do high discharge gas temperatures do to the oil in a compressor?

9. How can a hot hermetic compressor be cooled with water?

10. Which of the following terminal combinations would show an open circuit if a hermetic compressor with an internal overload were to be open due to a hot compressor?

    A. Common to run

    B. Common to start

    C. Run to start

Lab 31 Compressor Efficiency

Name ______________________________ Date ____________ Grade ________

OBJECTIVES: Upon completion of this exercise, you will be able to evaluate the performance of a hermetic compressor.

INTRODUCTION: You will use a gage manifold to measure the suction and head pressure at design conditions. You will compare this with the full-load amperage of the compressor to determine operating characteristics. To achieve design conditions, the system must be operated until the temperature in the conditioned space is lowered to near the design level.

TEXT REFERENCES: Unit 20.

TOOLS AND MATERIALS: A gage manifold, goggles, gloves, adjustable wrench, voltmeter, clamp-on ammeter, valve wrench, and a refrigeration system with known full-load amperage for the compressor motor.

SAFETY PRECAUTIONS: Wear goggles and gloves while working with refrigerant under pressure.

PROCEDURES

1. Put on your goggles and gloves. With the power off, tagged and locked out, fasten the gages to the valve ports or service valves and obtain a gage reading.
2. Fasten the ammeter to the common or run terminal leading to the compressor.
3. Start the system.
4. Allow the system to run until the conditioned space is near the design temperature, just before the thermostat would shut it off.
5. Lower the thermostat so it will not stop the compressor.
6. Reduce the airflow across the condenser until the refrigerant is condensing at 125°F. The head pressure will match one of the following:
   - 169 psig for an R–12 system
   - 278 psig for an R–22 system
   - 301 psig for an R–502 system
7. Record the following information:
   - Suction pressure __________psig
   - Boiling temperature __________°F
   - Discharge pressure __________psig
   - Condensing temperature __________°F
   - Compressor rated full load amperage __________A
   - Compressor actual amperage __________A
   - Compressor rated voltage __________V
   - Actual voltage at the unit __________V

8. If the compressor will pump the pressure differential, and the full-load amperage is within 10% of the rated amperage, the compressor is pumping to near capacity. It is doing all it can.

9. Remove the gages.

MAINTENANCE OF WORK STATION AND TOOLS: Return all tools to their respective places. Replace all panels. Return the unit to its normal running condition.

SUMMARY STATEMENT: How did the compressor in the above test compare to the rated pressures and rated amperage?

QUESTIONS

1. Name three types of compressors.

2. Name two different types of compressor drives.

3. Name two ways that small compressor motors are cooled.

4. Why should you never front seat the discharge service valve while a compressor is running?

5. What does the amperage of a compressor do when the head pressure rises?

6. What would the symptoms be for a compressor that has two cylinders and only one is pumping?

7. Is the full-load amperage printed on all compressors?

8. What lubricates the compressor's moving parts?

LAB 32 Hermetic Compressor Changeout

Name ______________________________ Date ____________ Grade ________

OBJECTIVES: Upon completion of this exercise, you will be able to change a hermetic compressor in a refrigeration system.

INTRODUCTION: You will remove the refrigerant charge, completely remove a compressor from a refrigeration system, and then replace it in the system. You will then leak check, evacuate, and charge the system as though the compressor were a new one.

TEXT REFERENCES: Unit 20.

TOOLS AND MATERIALS: A torch arrangement, leak detector (soap bubbles or halide will do), tubing cutters, tube reamer, gage manifold, goggles, gloves, line tap valves (if needed, such as for a small system), flat blade and Phillips screwdrivers, small socket set, refrigerant recovery system, scales or charging cylinder may be needed in order to add the correct charge, vacuum pump, and a cylinder of refrigerant.

SAFETY PRECAUTIONS: Wear goggles and gloves while transferring liquid refrigerant. DO NOT VENT REFRIGERANT TO THE ATMOSPHERE.

PROCEDURES

1. Put on your goggles and gloves. Fasten gages to the system and purge the gage lines.
2. Remove the refrigerant from the system to the appropriate standard using a refrigerant recovery system.
3. Turn off and lockout the power, remove the wiring from the compressor terminals, and mark each wire.
4. Remove the compressor hold down nuts or bolts.
5. Use tubing cutters to cut the compressor out of the system. Cut the compressor suction and discharge lines at least 2" from the shell so you have a stub of piping to work with when replacing the compressor.
6. Remove the compressor from the appliance or frame and set it aside.
7. Ream the compressor stubs and the piping in the system of the burrs. DO NOT ALLOW ANY MATERIAL TO FALL DOWN THE PIPE INTO THE SYSTEM OR THE COMPRESSOR.
8. Now replace the compressor onto its mountings, as though it were a new one. NOTE: REPLACE THE COMPRESSOR AS FAST AS POSSIBLE AND YOU WILL NOT NEED TO REPLACE THE LIQUID-LINE DRIER.
9. Use sweat tubing connectors to fasten the suction and the discharge lines back together. A good grade of low temperature solder may be what your instructor would recommend so this process may be performed again on the same unit.
10. Pressure the system to cylinder pressure and leak check the connections you worked with.
11. Triple evacuate the system. The motor terminal wiring and compressor mounting bolts may be refastened at this time.
12. Charge the system with the correct amount of refrigerant. Scales or a charging cylinder are recommended for critically charged systems.
13. Return the system to the condition your instructor directs you.

MAINTENANCE OF WORK STATION AND TOOLS: Return all tools to their correct locations.

SUMMARY STATEMENT: Describe when a new filter drier would be vital for this system.

QUESTIONS

1. How long did it take you to triple evacuate the system?

2. What type of service ports did you use on this system?

3. How long did it take to remove the refrigerant from the system?

4. Why is it not recommended that the refrigerant be exhausted to the atmosphere?

5. Draw the terminal designation for the compressor, showing common, run, and start.

6. Was this compressor mounted on external or internal springs?

7. How much refrigerant charge was used to recharge the unit?

LAB 33 Condenser Functions and Superheat

Name ______________________________ Date ______________ Grade ________

OBJECTIVES: Upon completion of this exercise, you will be able to take refrigerant and air temperature readings at the condenser. You will be able to determine condensing temperatures from the pressure-temperature chart using the compressor discharge pressure, and determine the condensing superheat under normal and loaded conditions.

INTRODUCTION: You will apply a manifold gage and thermometers to an air-cooled condenser of a typical refrigeration system to take pressure and temperature readings. You will also determine the superheat at the condenser under normal and loaded conditions.

TEXT REFERENCES: Unit 21.

TOOLS AND MATERIALS: A thermometer that reads up to 200°F and that can be strapped to the compressor discharge line, small piece of insulation, thermometer for recording the condenser air inlet temperature, goggles, light gloves, manifold gage, piece of cardboard, straight blade screwdriver, 1/4" and 5/16" nut driver, and a pressure-temperature conversion chart.

SAFETY PRECAUTIONS: The discharge line may be hot, so give it time to cool with the compressor off before fastening the thermometer. Be very careful when working around the turning fan. Use care in fastening the refrigerant gages. Wear goggles and light gloves. Do not proceed with this exercise until your instructor has given you instruction in the use of the manifold gage.

PROCEDURES

1. With the compressor off, and after the discharge line has cooled, fasten a thermometer to the discharge line, insulate it, and place a thermometer in the air entering the condenser. See Figure 1.

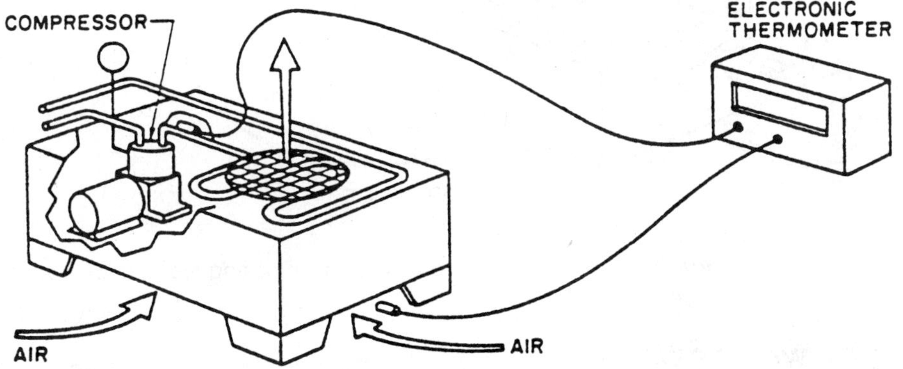

Figure 1 Thermometer loads to compressor discharge line and where airflow enters condenser.

2. Put on your goggles and gloves. Fasten the high-pressure gage to the high-pressure side of the system. You may also fasten a low-side gage if you want, although you will not need it.

3. Start the unit and record the following after 15 minutes.

   Refrigerant type: R-_____

   Discharge pressure: __________psig

   Discharge line temperature: __________°F

Condensing temperature (converted from pressure on the PT chart): __________°F

Discharge line superheat (discharge line temperature – condensing temperature): __________°F

Air temperature entering the condenser: __________°F

Temperature difference (condensing temperature – air temperature): __________°F

4. By blocking the condenser airflow, Figure 2, increase the head (discharge) pressure by 25 psig and record the following:

   Discharge pressure: __________psi

   Discharge line temperature: __________°F

   Condensing temperature (converted from pressure on the PT chart): __________°F

   Discharge line superheat (discharge line temperature – condensing temperature): __________°F

   Air temperature entering the condenser: __________°F

   Temperature difference (condensing temperature – air temperature): __________°F

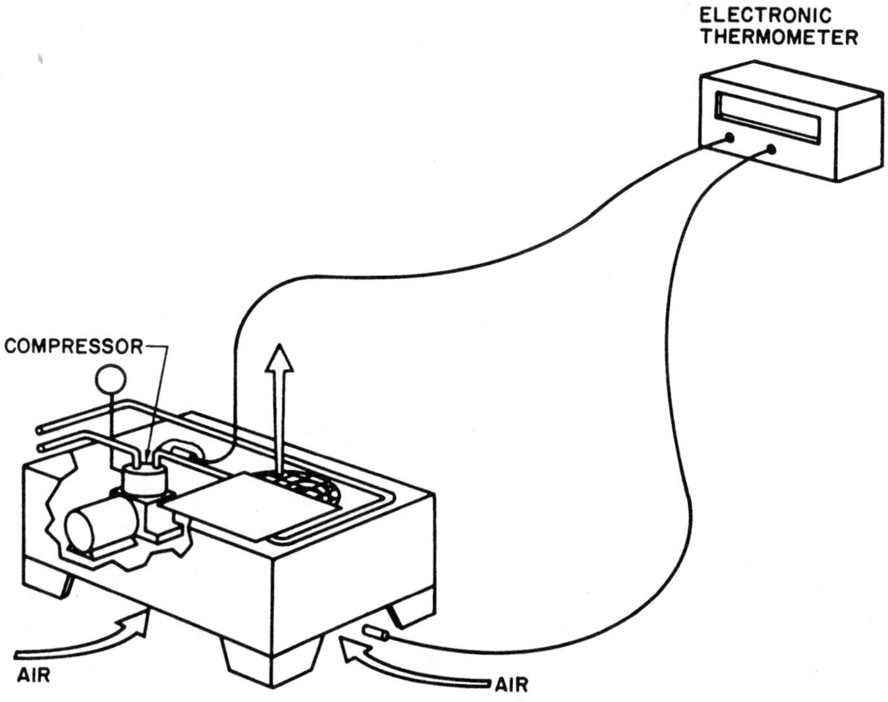

Figure 2 Increasing the head pressure by partially blocking the condenser.

MAINTENANCE OF WORK STATION AND TOOLS: Stop the unit and remove the thermometer from the discharge line after it has cooled, remove the gages and wipe off any oil that may have accumulated on the unit. Replace all tools. Replace all panels and leave the unit as you found it.

SUMMARY STATEMENT: Describe why the compressor discharge line is so much hotter than the condensing temperature and why it does not follow the temperature and pressure relationship.

QUESTIONS

1. What must be done to the hot gas leaving the compressor before it may be condensed?

2. The refrigerant leaving the condenser must be a __________. (liquid, vapor, or a solid)

3. What is a typical difference in temperature between the condensing refrigerant and the air entering the condenser?

4. What happens to the temperature of the discharge line when the head pressure is increased?

5. Where does the condensing take place in a typical condenser?

6. What would the head pressure do if the fan motor on an air-cooled condenser stopped working?

7. What job does the condenser do in the refrigeration system?

8. Where is the condenser located in relation to the evaporator on a typical refrigeration system?

9. The head pressure of the refrigerant in a running refrigeration machine is determined by the condensing temperature. (True or false)

10. The condensing temperature is determined by the condensing medium – air or water. (True or false)

LAB 34 Air-Cooled Condenser Performance

Name ______________________________ Date ____________ Grade ________

OBJECTIVES: Upon completion of this exercise, you will be able to perform basic evaluation procedures on an air-cooled condenser.

INTRODUCTION: You will use a thermometer and a gage manifold to evaluate an air-cooled condenser to determine its performance.

TEXT REFERENCES: Unit 21.

TOOLS AND MATERIALS: A gage manifold, thermometer, goggles, light gloves, adjustable wrench, valve wrench, and a refrigeration system with an air-cooled condenser and service valves.

SAFETY PRECAUTIONS: Wear goggles and gloves while connecting and disconnecting gage lines. Make sure you have had instruction in using a gage manifold.

PROCEDURES

1. Put on your goggles and gloves. Fasten the high- and low-side gage manifold lines to the service valves.
2. Turn the service valves off the back seat (clockwise) enough to obtain a gage reading.
3. Purge the gage lines to clean them by loosening the fittings slightly at the manifold.
4. Place the thermometer in the air stream entering the condenser.
5. Start the system and wait until it is within 10°F of the correct temperature range inside the box.
6. Record the following:
   - Type of box (low, medium, or high temperature) ________
   - Type of condenser (with or without fan) ________
   - Suction pressure ________psig
   - Head pressure ________psig
   - Condensing temperature (converted from head pressure) ________°F
   - Condensing entering air temperature ________°F
   - Temperature difference (condensing temperature minus air-in temperature) ________°F
7. Is this condenser running at less than 30°F temperature difference? (yes or no) If the answer is yes, it is probably running efficiently.
8. If the temperature difference is more than 30°F, check the condenser to see if it is clean.
9. Remove the gages.

MAINTENANCE OF WORK STATION AND TOOLS: Return the refrigeration system to its proper condition. Clean the work station and return all tools to their respective places.

SUMMARY STATEMENT: Describe the condenser-to-entering-air relationship of a typical air-cooled condenser.

QUESTIONS

1. Name the three functions that a condenser performs.

2. What is considered the highest temperature that the discharge gas should reach upon leaving the compressor?

3. What may be used on an air-cooled condenser with a fan to prevent the head pressure from becoming too low in mild weather?

4. What are the system results of an air-cooled condenser operating with a head pressure that is too low?

5. Why must an air-cooled condenser be located where the air will not recirculate through it?

6. What does the compressor amperage do when the head pressure rises?

7. What would the symptoms be if a condenser had two fan motors and one quit running?

8. What is the purpose of the fins on the condenser?

9. What materials are typical refrigeration condensers made of?

LAB 35 Evaluation of a Water-Cooled Condenser

Name ______________________________ Date ______________ Grade ________

OBJECTIVES: Upon completion of this exercise, you will be able to evaluate a simple water-cooled refrigeration system.

INTRODUCTION: You will use gages and a thermometer to evaluate a simple water-cooled system. You will do this by evaluating the heat exchange in the condenser.

TEXT REFERENCES: Unit 21.

TOOLS AND MATERIALS: A gage manifold, goggles, light gloves, adjustable wrench, valve wrench, thermometer, and a water-cooled refrigeration system.

SAFETY PRECAUTIONS: Wear goggles and gloves while working with refrigerant under pressure.

PROCEDURES

1. Put on your goggles and gloves. Connect the high- and low-side gages.
2. Slightly open the service valves from their back seat, far enough for the gages to register the pressure.
3. Fasten one thermometer lead to the discharge line of the compressor.
4. Place one other lead in the entering water and one in the leaving water of the condenser in such a manner that the lead will give accurate water readings. If there are no thermometer wells, you may use the method in the text.
5. Start the system and allow it to run until it is operating at a steady state, that is, until the readings are not fluctuating. Record the following:
   - Type of refrigerant __________
   - Suction pressure __________ psig
   - Discharge pressure __________ psig
   - Discharge line temperature __________°F
   - Condensing temperature (converted from discharge pressure) __________°F
   - Condenser leaving water temperature __________°F
   - Condenser entering water temperature __________°F
   - TD (temperature difference between the leaving water and the condensing refrigerant) __________°F
6. Disconnect the gages and thermometer leads. Replace all panels that may have been removed with the correct fasteners.

MAINTENANCE OF WORK STATION AND TOOLS: Return all tools to their respective places.

SUMMARY STATEMENTS: Describe the above system's ability to transfer heat into the water. Was the system transferring the heat at the correct rate?

QUESTIONS

1. Is a water-cooled condenser more efficient than an air-cooled condenser?

2. Why are water-cooled condensers not as popular as air-cooled condensers in small equipment?

3. What is the typical difference in temperature between the condensing refrigerant and the leaving water for a condenser that is operating efficiently?

4. Name two places that the heat-laden water from a water-cooled condenser may be piped to.

5. Why is the discharge line hotter than the condensing refrigerant?

6. Name the three functions that take place in the condenser.

7. What controls the waterflow through a water regulating valve?

8. What does the temperature difference across a condenser do if the waterflow is reduced?

9. What does the temperature difference between the condensing refrigerant and the leaving water do if the condenser tubes are dirty?

10. What is the low form of plant life that grows in condenser water?

LAB 36 Determining the Correct Head Pressure

Name ______________________________ Date ______________ Grade ________

OBJECTIVES: Upon completion of this exercise, you will be able to determine the correct head pressure for a system under the existing operating conditions for air-cooled condensers.

INTRODUCTION: You will use pressure and temperature measurements to determine the correct head pressure for an air-cooled condenser under typical operating conditions. If a unit has an abnormal load on the evaporator, the head pressure will not conform exactly to the ambient temperature relationship.

TEXT REFERENCES: Unit 21.

TOOLS AND MATERIALS: A gage manifold, goggles, light gloves, adjustable wrench, straight blade and Phillips screwdrivers, 1/4" and 5/16" nut drivers, thermometer, and an air-cooled refrigeration system with service valves or Schrader valve ports.

SAFETY PRECAUTIONS: Wear goggles and gloves while working with refrigerant under pressure.

PROCEDURES

1. Your refrigeration system should have been running for a period of time so that it is operating under typical conditions. Using the thermometer, determine the temperature of the air entering the condenser. Record here: __________°F
2. Determine from the air entering the condenser what you think the maximum condensing temperature should be. Air entering + 30°F maximum = __________°F
3. Convert the above condensing temperature to head pressure using the pressure/temperature relationship chart in the text. __________ psig
4. Put on your goggles and gloves. Now that you have determined what the head pressure should be, connect the high- and low-side gage lines and record the following:
   - Actual head pressure __________ psig
   - Actual condensing temperature __________°F
   - Air temperature entering condenser __________°F
   - TD (condensing temperature minus temperature of the air entering the condenser) __________°F

   If the last item is over 30°F, is the system under excess load? Is the evaporator absorbing excess heat because it is operating above its design application?
5. Remove the gages and replace all panels with the correct fasteners.

MAINTENANCE OF WORK STATION AND TOOLS: Return all tools to their respective places.

SUMMARY STATEMENTS: Describe the action in the condenser that you have been working with. Was it performing the way it should under the existing conditions?

QUESTIONS

1. State the three functions of a condenser.

2. Why does the head pressure rise when the evaporator has an excess load?

3. What would the TD be for a high-efficiency condenser?

4. What two temperatures are used for TD on a water-cooled condenser?

5. What TD is used to determine the condenser performance of a water-cooled condenser?

6. Why must a minimum head pressure be maintained for equipment, particularly if it is located outside?

7. Name three methods used to control head pressure on air-cooled equipment.

8. What method is used to control head pressure on water-cooled equipment that wastes the water?

9. Why would the entering water temperature vary on a waste water system?

10. What would be the symptoms for an air-cooled condenser where the air leaving the condenser is hitting a barrier and recirculating?

LAB 37 Expansion Devices

Name ______________________ Date ____________ Grade ________

OBJECTIVES: Upon completion of this exercise, you will be able to describe the characteristics of a thermostatic expansion valve and adjust it for more or less superheat.

INTRODUCTION: You will use a system with an adjustable thermostatic expansion valve and check the superheat when the valve is set correctly. You will then make adjustments to produce more and less superheat and watch the valve respond to adjustment.

TEXT REFERENCES: Unit 22.

TOOLS AND MATERIALS: A gage manifold, thermometer, goggles, gloves, adjustable wrench, valve wrench, and a refrigeration system with a thermostatic expansion valve.

SAFETY PRECAUTIONS: Wear goggles and gloves while connecting gages to the system.

PROCEDURES

1. Put on your goggles and gloves. Connect the gage manifold lines to the high- and the low-side of the system.
2. Fasten the thermometer sensor to the suction line just as it leaves the evaporator. If there is a heat exchanger, make sure that the sensor is fastened between the evaporator and the heat exchanger.
3. Inspect the thermostatic expansion valve sensing bulb and be sure that it is fastened correctly to the line. It should be insulated.
4. Turn the thermostat to a low setting where it will not satisfy and shut the system off.
5. Start the system and wait until it is operating at very close to design conditions (high, medium, or low temperature).
6. Record the following data when the system is operating in a stable manner:
   - Discharge pressure __________ psig
   - Condensing temperature __________°F
   - Suction line temperature __________°F
   - Suction pressure __________ psig
   - Boiling temperature __________°F
   - Superheat __________°F
7. Turn the thermostatic expansion valve adjustment two full turns for an increase in superheat and wait for the system to stabilize. Record the following data:
   - Discharge pressure __________ psig
   - Condensing temperature __________°F
   - Suction line temperature __________°F
   - Suction pressure __________ psig
   - Boiling temperature __________°F
   - Superheat __________°F

8. Turn the valve adjustment back to where it was and then two full turns toward a decrease in superheat and wait for the system to stabilize. Record the following:
   - Discharge pressure __________ psig
   - Condensing temperature __________°F
   - Suction line temperature __________°F
   - Suction pressure __________ psig
   - Boiling temperature __________°F
   - Superheat __________°F
9. Turn the valve adjustment back to the mid-position where it was in the beginning, and verify that the superheat is maintaining between 8°F and 12°F.
10. Remove the gages and thermometer lead.

MAINTENANCE OF WORK STATION AND TOOLS: Return all tools and equipment to their respective places. Be sure to replace all panels and fasteners in the proper manner. Leave the system in the manner you found it. Be sure to turn the thermostat back to normal.

SUMMARY STATEMENT: Describe the change in superheat when adjusted. Indicate how much change per turn was experienced.

QUESTIONS

1. When a coil is starved for refrigerant, the superheat is __________.
2. When a coil is flooded with refrigerant, the superheat is __________.
3. The thermostatic expansion valve maintains a constant __________.
4. The automatic expansion valve maintains a constant __________.
5. The thermostatic expansion valve feeds (more or less) refrigerant with an increase in load.
6. The automatic expansion valve feeds (more or less) refrigerant with an increase in load.

LAB 38 Troubleshooting Exercise, Air Conditioner with TXV

Name ______________________________ Date ______________ Grade __________

OBJECTIVES: Upon completion of this exercise, you will be able to evaluate an air conditioning system and make a repair on a no-cooling problem.

INTRODUCTION: This air conditioning unit will have a specified problem introduced to it by your instructor. You will locate the problem using a logical approach and instruments. THE INTENT IS NOT FOR YOU TO MAKE THE REPAIR, BUT TO LOCATE THE PROBLEM AND ASK YOUR INSTRUCTOR FOR DIRECTION BEFORE COMPLETION.

TEXT REFERENCES: Unit 22.

TOOLS AND MATERIALS: A VOM, ammeter, gage manifold, thermometer, goggles, gloves, straight blade and Phillips screwdrivers, electrical tape, short piece of insulation, 1/4" and 5/16" nut drivers.

SAFETY PRECAUTIONS: Use care while touch testing the unit. Do not allow your hands close to the fan blades. DO NOT JUMP OUT ANY HIGH PRESSURE CONTROL OR OVERLOAD. Wear goggles and gloves when connecting gage lines.

PROCEDURES

1. Put on your goggles and gloves. Connect a gage manifold and purge the gage lines.
2. Start the unit. If the unit will not run, use the VOM to determine why. It may be off because of temperature, low pressure or high pressure. If the unit is off because of low pressure, you may jump the low pressure control for a short moment only to see if the unit will start. Gages should be watched at this time. Do not allow the low-side pressure to go into a vacuum.
3. With the unit running, touch test the suction line at the compressor. NOTE: The unit must have been running for at least 15 minutes. How does the suction line feel in relation to your hand temperature? Is it warmer or cooler?
4. Carefully touch the compressor discharge line. How does it feel? Is it hot or warm?
5. Touch test the suction line leaving the evaporator. Does it feel like the line entering the compressor? (yes or no)
6. Touch test the line leaving the TXV expansion valve. How does it feel? _________
7. What conclusions did you draw from the above touch test?
8. Is liquid refrigerant moving in the liquid line sight glass? (yes or no)
9. Describe the problem and your recommended repair.

MAINTENANCE OF WORK STATION AND TOOLS: Return all tools and equipment to their respective places.

SUMMARY STATEMENT: Describe how you decided whether you had an electrical or mechanical problem.

QUESTIONS

1. Describe a starved evaporator.

2. What should the superheat be at the end of the evaporator in a typical system?

3. Why is superheat vital at the end of the evaporator?

4. What would the symptoms of too much superheat be?

5. What could the problem be when the liquid line sight glass is full of liquid at the expansion valve, but only a small amount of liquid will feed through the valve?

6. How can a restriction occur in an expansion valve?

7. What is the usual symptom when the power element of a TXV loses its charge?

8. What would be the symptom if the TXV sensing element were not fastened correctly to the suction line?

9. Why should you allow the unit to run for 15 minutes before drawing conclusions?

10. What does it mean when the compressor discharge line is not hot?

LAB 39 Capillary Tube System Evaluation

Name ______________________________ Date ______________ Grade __________

OBJECTIVES: Upon completion of this exercise, you will be able to evaluate a system using a capillary tube as a metering device.

INTRODUCTION: You will use a thermometer and gage manifold to check the evaporator and capillary tube performance under light load and full load conditions for the correct charge under these changing conditions.

TEXT REFERENCES: Unit 22.

TOOLS AND MATERIALS: A gage manifold, thermometer, goggles, gloves, straight blade and Phillips screwdrivers, electrical tape, short piece of insulation, 1/4" and 5/16" nut drivers, and a refrigeration system with service valves.

SAFETY PRECAUTIONS: Wear goggles and gloves while working with refrigerant under pressure.

PROCEDURES

1. Put on your goggles and gloves. Connect the high- and low-side gage lines to the unit. The connections will probably be Schrader-type connectors for a capillary tube system.
2. Purge a small amount of refrigerant from each gage line to clean the gages of contaminants.
3. Fasten a thermometer lead to the suction line at the end of the evaporator, but before any heat exchanger in the line.
4. Insulate the thermometer lead.
5. Start the system and let it run for 15 minutes and record the following information:
   - Suction pressure __________ psig
   - Suction line temperature __________°F
   - Refrigerant boiling temperature (converted from suction pressure) __________°F
   - Superheat (subtract boiling temperature from line temperature) __________°F
   - Discharge pressure __________ psig
   - Refrigerant condensing temperature (converted from discharge pressure) __________°F
6. Block the condenser airflow until the condensing temperature is 125°F and the discharge pressure reads one of the following:
   - R-12 169 psig
   - R-22 278 psig
   - R-502 301 psig

7. Allow the system to run for at least 15 minutes before taking the next readings. You may have to turn the thermostat down to a lower setting to prevent the system from stopping. Take the following readings:
   - Suction pressure __________ psig
   - Refrigerant boiling temperature __________ °F
   - Suction line temperature __________ °F
   - Discharge pressure __________ psig
   - Refrigerant condensing temperature __________ °F
   - Superheat (subtract boiling temperature from line temperature) __________ °F
8. Shut the unit off and remove the gages. Replace all panels with the correct fasteners.

MAINTENANCE OF WORK STATION AND TOOLS: Return all tools to their respective places. Insure that your work area is left clean.

SUMMARY STATEMENTS: Describe what the superheat in the evaporator did when the head pressure was raised to the condensing temperature of 125 °F. Explain why.

QUESTIONS

1. Why is a system using a capillary tube thought of as a dry-type or direct expansion system?

2. Would an orifice-type expansion device perform much the same as a capillary tube metering device?

3. What causes refrigerant to flow through a capillary tube metering device?

4. What are the symptoms of a capillary tube metering device system with an overcharge of refrigerant?

5. What are the symptoms of a capillary tube metering device system with an undercharge of refrigerant?

6. Name two advantages of a capillary tube metering device.

LAB 40 Changing a Thermostatic Expansion Valve

Name ______________________________ Date ______________ Grade ________

OBJECTIVES: Upon completion of this exercise, you will be able to change a thermostatic expansion valve with a minimum loss of refrigerant.

INTRODUCTION: You will change the TXV on a working unit. Because there may not be a unit with a defective valve, you will go through the exact change out procedure by removing the valve and then replacing it. The system should have service valves and a TXV that is installed with flare fittings.

TEXT REFERENCES: Unit 22.

TOOLS AND MATERIALS: A gage manifold, goggles, gloves, a refrigerant-recovery system, a cylinder of the same type of refrigerant as is in the system, leak detector (soap or halide will do) service valve wrench, two 8" adjustable wrenches, assortment of flare nut wrenches, a vacuum pump, flat blade and Phillips screwdrivers.

SAFETY PRECAUTIONS: Watch the high pressure gage while pumping the system down to be sure the condenser will hold the complete charge. You may need to jump the low pressure control out to pump the unit down. DO NOT JUMP OUT THE HIGH PRESSURE CONTROL.

PROCEDURES

1. Put on your goggles and gloves. Fasten gages to the compressor service valves and purge the gage manifold.
2. Front seat the "king" valve at the receiver, or close the liquid-line service valve.
3. Jump the low-pressure control if there is one.
4. Start the unit and allow it to run until the low-side pressure drops to 0 psig. Turn the unit off and allow it to stand for a few minutes. If the pressure rises, pump it down again. You will normally need to pump the system down to 0 psig 3 times in order to remove all refrigerant from the low side. Remember, you are pumping the drier and the liquid line down also.
5. Remove enough panels from the evaporator to comfortably reach the TXV valve.
6. Remove the TXV sensor from the suction line.
7. Using the adjustable wrenches, or the flare nut wrenches, loosen the inlet and outlet connections to the TXV. Then loosen the external equalizer line. Remove the valve from the system.
8. When the valve is completely clear of the evaporator compartment, you may then replace it as though the valve you removed were a new one. THIS SHOULD BE DONE QUICKLY SO THE SYSTEM IS NOT OPEN ANY LONGER THAN ABSOLUTELY NECESSARY, OR THE DRIER SHOULD BE CHANGED.
9. When the valve is replaced, pressure the system up to cylinder pressure and leak check.
10. Recover the refrigerant used for leak detection and connect the vacuum pump. Triple evacuate, using the procedures in Unit 5.
11. After evacuation, open the "king" valve and start the system. The charge should be correct.
12. Return the unit to the condition your instructor advises.

MAINTENANCE OF WORK STATION AND TOOLS: Return all tools to their correct places.

SUMMARY STATEMENTS: Describe the triple evacuation process.

QUESTIONS

1. How did you change the TXV and get by without changing the filter drier?

2. Why is triple evacuation better than one short vacuum?

3. Why should you NEVER jump out the high-pressure control?

4. What would be the symptom of a TXV that had lost its charge in the sensing element?

5. What is the advantage of a flare nut wrench over an adjustable wrench?

6. How was the sensing element fastened to the suction line on the unit you worked on?

7. Was the valve sensing bulb insulated?

8. What is the purpose of the insulation on the sensing bulb?

9. Is it necessary that all sensing bulbs be insulated?

10. Did this unit have a low-pressure control?

LAB 41 Evaluating Duct Size

Name ______________________________ Date ______________ Grade __________

OBJECTIVES: Upon completion of this exercise, you will be able to use a duct chart to evaluate the duct size on a simple residential or commercial duct system for adequate airflow in heating or cooling cycles.

INTRODUCTION: You will evaluate a simple residential duct system using the duct evaluation chart in the text.

TEXT REFERENCES: Unit 25.

TOOLS AND MATERIALS: A flashlight, straight blade and Phillips screwdrivers, 1/4" and 5/16" nut drivers, and a tape measure.

SAFETY PRECAUTIONS: Care should be taken not to cut your hands on sheet metal edges while taking duct measurements.

PROCEDURES

1. You will be assigned a duct system to evaluate. Find the capacity of the system and record here: heating __________ Btuh, cooling __________ Btuh.
2. Record the number of supply outlets. __________
3. Record the number of return outlets.__________
4. Size of the supply duct __________ in. Return inlet __________ in.
5. Recommended size from chart, supply __________ in. Return __________ in.
6. Recommended number and size of supply outlets __________
7. Recommended number of return inlets __________
8. Fan wheel size, width __________ in. Diameter __________ in.
9. Type of fan motor, direct-drive or belt-drive
10. Draw a line diagram of the duct system you are working with, showing the sizes of the supply and return trunk and the individual branch ducts.

MAINTENANCE OF WORK STATION AND TOOLS: Replace all panels with the correct fasteners.

SUMMARY STATEMENT: Describe the symptoms of a gas furnace that does not have adequate airflow.

QUESTIONS

1. Describe static pressure.

2. Describe velocity pressure.

3. Describe total pressure.

4. Name a common gage for measuring static pressure.

5. What unit of pressure is commonly used to describe air pressure in a duct system?

6. Would the static pressure be a positive or negative value in the supply duct?

7. Would the static pressure be a positive or negative value in the return duct?

8. If the velocity of air in a 12 × 20 inch supply duct is 700 feet per minute, how much air is moving? ________ cfm

LAB 42 Airflow Measurements

Name ______________________ Date ____________ Grade ________

OBJECTIVES: Upon completion of this exercise, you will be able to use basic airflow measuring instruments to measure airflow from registers and grilles.

INTRODUCTION: You will use a prop-type velometer such as in Figure 1 in the text to record the velocity of the air leaving the register. You will then measure the size of the register or use the manufacturer's data to determine in cfm the volume of air leaving the register.

TEXT REFERENCES: Unit 25.

TOOLS AND MATERIALS: A prop-type velometer or the equivalent measuring device, and goggles.

SAFETY PRECAUTIONS: Do not allow air to blow directly into your eyes, because it may contain dust.

Figure 1 Prop-type velometer.
Photo by Bill Johnson

PROCEDURES

1. Take your air velocity measurements from a rectangular air discharge register that has a flat face, such as a high-side wall, floor, or ceiling type.
2. Start the system fan.
3. Measure the velocity in several places on the register for an average reading. Record the readings here in FPM, feet per minute.

- ________, ________, ________, ________, ________
- Average the above readings and record here. ________
- The volume of air may be found in the manufacturer's literature.
- The volume of air in cfm may be found by multiplying the velocity of the air in feet per minute (FPM) times the free area of the register in square feet. The free area of the register may be found by the following method:

  Measure the open area of the front of the register where the air flows out, the fin area. For example, the register may measure 3.5" × 5.5" = 19.25 square inches. Divide by 144 square inches to get the number of square feet (0.134 square feet area register).
  - The area in the above step represents the face area, not the free area.
  - The metal blades take up some of the area that you measured. For a register with movable blades, multiply the area × 0.75 (75% free area).

  For a register with stamped blades, multiply the area × 0.60 (60% free area).

- ________ FPM × ________ square feet = ________ CFM.

4. You may take the above measurements on each outlet in the system and add together for the total airflow in the system.

5. If the system has one or two return-air inlets, you may arrive at the total airflow by using the above method with these grilles. Remember, the air is flowing toward the return-air grille.

MAINTENANCE OF WORK STATION AND TOOLS: Return all tools to their respective places.

SUMMARY STATEMENT: Describe what free area in a register or grille means.

QUESTIONS

1. How is a stamped grille different from a register?

2. Which of the above is the most desirable for directing airflow?

3. Which of the above is the most expensive?

4. What is the difference between a register and a grille?

5. What is the typical static pressure behind a supply register that forces air into the room?

6. What problem would excess velocity from the supply registers cause?

7. What would the symptoms be if supply registers (that blow air across the floor) designed for heating were used for air conditioning?

8. A duct that is 12" × 24" has an air velocity of 800 fpm. What is the airflow in cfm in the duct?

9. A supply register with movable blades has a face measurement of 6.5" × 12.5". How much air is flowing out of the register if the average face velocity is 300 fpm? Show your work.

LAB 43 Evaluating an Air Conditioning Installation

Name ______________________________ Date ______________ Grade ________

OBJECTIVES: Upon completion of this exercise, you will be able to look at an air conditioning system and evaluate the installation for good workmanship and completeness.

INTRODUCTION: You will use this lab exercise as a checksheet to insure a split-system air conditioning installation is complete and ready for startup.

TEXT REFERENCES: Unit 26.

TOOLS AND MATERIALS: Straight blade and Phillips screwdrivers, VOM, clamp-on ammeter, 1/4" and 5/16" nut drivers, tape measure, and a flashlight.

SAFETY PRECAUTIONS: Turn the power off and verify with the VOM that it is off. Lock and tag the panel or disconnect box where the power is turned off. You should have a single key. Keep it in your possession while the power is off.

PROCEDURES

Use the following checklist to insure the installation is in good order. Turn off and lock out the power.

1. Evaporator section:

   Air handler level _____, vibration eliminator installed _____, duct fastened tightly _____, connections taped _____, filter in place _____, auxiliary drain pan if evaporator in attic _____, secondary condensate drain line piped correctly _____, primary drain pan piped correctly and trap with clean-out plug _____, condensate pump, if needed, to pump the condensate to the drain _____, electrical connections in the air handler tight, high voltage correct _____ low voltage correct _____, fan motor and wheel can be removed easily _____

2. Condenser section:

   Condenser level _____, a firm foundation under the condenser _____, service panel in correct location _____, airflow correct and not recirculating _____, refrigerant piping in good order and insulated _____, electrical connections tight _____, high voltage correct _____, low voltage correct _____, service valves open, if any _____

3. Condensate drain termination:

   In a dry well _____, on the ground _____, in a drain _____

4. Voltage check:

   Line voltage to indoor fan section _________ V, to condenser _________ V, Low-voltage power supply voltage _________ V

5. Start the indoor fan motor and verify the voltage and current: Rated voltage _________ V, actual voltage _________ V, rated current _________ A, actual current _________ A

6. Start the condensing unit and verify the voltage and amperage: Rated voltage _________ V, actual voltage _________ V, compressor rated current _________ A, actual current _________ A, fan motor rated current _________ A, actual current _________ A

7. Replace all panels with the correct fasteners.

MAINTENANCE OF WORK STATION AND TOOLS: Return all tools to their places.

SUMMARY STATEMENTS: Describe the installation you inspected as it compares to the checklist. Describe any improvements you would have made.

QUESTIONS

1. What are vibration isolators for?

2. Describe a dry well for condensate.

3. What is the purpose of insulation on the suction line?

4. What is the purpose of an auxiliary drain pan when the evaporator is in the attic?

5. What is the purpose of the trap in the condensate drain line?

6. Why should water from the roof not drain into the top of a top discharge condenser?

7. How can an evaporator section be suspended from floor joists in a crawl space?

8. Why should air not be allowed to recirculate back into a condenser?

9. What would the symptoms of reduced airflow be at the evaporator?

10. What should be done if the voltage to the unit is too low?

LAB 44 Using a Halide Leak Detector

Name ______________________________ Date ______________ Grade ________

OBJECTIVES: Upon completion of this exercise, you will be able to use a halide torch to leak-check a system for refrigerant leaks.

INTRODUCTION: With gages you will check to determine that there is refrigerant pressure in an air conditioning system. You will then check all field and factory connections with a halide leak detector to determine whether or not there are leaks. If the leak is from a flare connection, tighten it. If it is from a solder connection, ask your instructor what you should do.

TEXT REFERENCES: Unit 26.

TOOLS AND MATERIALS: A halide leak detector, a gage manifold, a VOM, straight blade and Phillips screwdrivers, 1/4" and 5/16" nut drivers, goggles, and light gloves. If a leak is found, you will need additional tools to repair it.

SAFETY PRECAUTIONS: Turn off the electrical power. Lock and tag the disconnect boxes to both the indoor and outdoor units before beginning the exercise. Check with the VOM to insure that the power to the system is off. Wear goggles and light gloves when attaching or removing gages. Beware of toxic fumes from halide torch.

PROCEDURES

You will need instruction in the use of the halide leak detector before beginning this exercise. Beware of toxic fumes.

1. With the power off and locked out and with goggles and gloves on, fasten the gage lines to the high- and low-side gage ports. Check the gage readings to insure that there is pressure.
2. With the VOM, check the electrical disconnect to both units to insure that the electrical power is off.
3. With the halide leak detector, check all joints (solder and flare) in the indoor and outdoor piping. Check the joints under the insulation by making a small hole in the insulation with a small screwdriver and holding the sensing tube to the hole.
4. Remove any additional necessary panels and check all factory connections.
5. If a leak is found in a flare joint, tighten it. If one is found in a solder joint, ask for instructions.
6. When you have established that there is no leak, or when you have finished, remove the gages and replace the panels.

MAINTENANCE OF WORK STATION AND TOOLS: Return all tools, equipment, and materials to their proper storage places. Clean your work station.

SUMMARY STATEMENTS: Describe the halide leak-check procedure.

QUESTION

1. What pulls the refrigerant to the sensor in a halide leak detector?

2. What is the normal color of the gas flame in a halide leak detector?

3. What is the color of the gas flame in a halide leak detector when refrigerant is present?

4. Can a halide leak detector be used to detect natural gas? Why?

5. What does the term halide refer to?

6. Name two gases that are commonly used to produce the flame in a halide leak detector.

7. Describe how a halide leak detector discovers leaks.

8. Does the halide leak detector work best in the light or in the dark?

LAB 45 Using Electronic Leak Detectors

Name ______________________________ Date ______________ Grade ________

OBJECTIVES: Upon completion of this exercise, you will be able to use an electronic leak detector to find refrigerant leaks.

INTRODUCTION: You will check an air conditioning system with gages to determine that there is refrigerant pressure. You will then check all field and factory connections with an electronic leak detector to determine whether or not there are refrigerant leaks.

TEXT REFERENCES: Unit 26.

TOOLS AND MATERIALS: An electronic leak detector, gage manifold, VOM, straight blade and Phillips screwdrivers, 1/4" and 5/16" nut drivers, goggles, and light gloves. If a leak is found, you will need additional tools to repair it.

SAFETY PRECAUTIONS: Turn off the electrical power to both the indoor and outdoor unit before beginning this exercise. Lock and tag the disconnect box and keep the key in your possession. Check with the VOM to ensure that the system is off. Wear goggles and light gloves when attaching or removing gages. Do not allow liquid refrigerant on your skin.

PROCEDURES

You will need instruction in the use of the electronic leak detector before beginning this exercise.

1. With the power off and locked out and with goggles and gloves on, fasten the gage lines to the high- and low-side gage ports. Check the gage readings to insure that there is pressure. If there is no pressure, check with your instructor for the next step.
2. With the VOM, check the electrical disconnect to both units to insure that the power is off.
3. Using the electronic leak detector check all connections (flare and solder) at the indoor unit. Check under fittings.
4. Check all connections at the outdoor unit.
5. Remove necessary panels and check the factory connections.
6. Use a small screwdriver to make a small hole in the insulation on the tubing at each joint, and insert the probe in the holes to check for leaks.
7. Tighten any flare connection where you have found a leak. Ask for instructions before repairing solder connections.
8. When you have finished, carefully remove the gage lines and replace all panels with proper fasteners.

MAINTENANCE OF WORK STATION AND TOOLS: Place all tools, equipment, and materials in their proper storage places. Insure that your work station is left clean.

SUMMARY STATEMENTS: Describe how the electronic leak detector operates. What are some of its good and bad points?

QUESTIONS

1. Can natural gas be checked with an electronic leak detector?

2. How can the refrigerant be moved from the point of the leak to the detector sensor?

3. What can be done to keep wind from affecting the leak detection?

4. What is the power source for an electronic leak detector?

5. How can soap bubbles be used to further pinpoint the leak when it is detected?

6. What is the purpose of the filter under the tip of the sensor on the detector?

7. Does the electronic leak detector work best in a light or dark place?

LAB 46 Startup and Checkout of an Air Conditioning System

Name ______________________________ Date ____________ Grade ________

OBJECTIVES: Upon completion of this exercise, you will be able to check out components of an air conditioning system for an orderly system startup.

INTRODUCTION: You will start up an air conditioning system, one component at a time, and check each one to insure that it is operating correctly.

TEXT REFERENCES: Unit 26.

TOOLS AND MATERIALS: A gage manifold, VOM, clamp-on ammeter, thermometer, straight blade and Phillips screwdrivers, 1/4" and 5/16" nut drivers, flashlight, electrical tape, light gloves, and goggles.

SAFETY PRECAUTIONS: Care should be taken while taking voltage and amperage readings on live circuits. Wear gloves and goggles while attaching and removing gages and taking pressure readings. When the power is turned off, lock and tag the panel.

PROCEDURES

With the unit off:

1. Check the line voltage to the indoor and outdoor units: Indoor unit ________ V, outdoor unit ________ V.
2. Record the rated current of the indoor fan ________ A, outdoor fan ________ A, and compressor ________ A.
3. With goggles and gloves on, fasten the gage lines to the condensing unit gage ports and put one temperature lead in the return air and one in the outdoor air to record the ambient temperature.
4. Disconnect the common wire going to the compressor and tape the end to insulate it.
5. Set the thermostat to the "off" position and turn the power to the fan section and the condenser section on.
6. Turn the fan switch to "on" and start the indoor fan motor. Check and record the voltage and amperage of the motor ________ V, ________ A. Compare with manufacturer's specifications.
7. Verify that air is coming out all registers and that they are open.
8. Turn the thermostat to call for cooling and check the voltage and amperage of the fan motor at the condensing unit ________ V, ________ A. Compare with the manufacturer's specifications.
9. Turn the power off and reconnect the compressor wire.
10. Turn the power on and the compressor will start. Record the voltage and current ________ V, ________ A. Compare with manufacturer's specifications.
11. Let the unit run and record the following information:
    - Indoor air temperature ________°F, outdoor ________°F
    - Suction pressure ________ psig, discharge pressure ________ psig
12. Carefully remove the gages.
13. Replace all panels with the correct fasteners.

MAINTENANCE OF WORK STATION AND TOOLS: Return all tools to their correct places. Leave your

MAINTENANCE OF WORK STATION AND TOOLS: Return all tools to their correct places. Leave your work station clean.

SUMMARY STATEMENT: Compare the temperature for the suction pressure on the above system in relation to the inlet air temperature. Indicate whether or not the head pressure was correct for the conditions.

QUESTIONS

1. Why is it wise to disconnect the compressor while going through a startup?

2. What should be done if the line voltage is too high before startup?

3. What should be done if the line voltage is too low before startup?

4. How does the outdoor temperature affect the head pressure?

5. How does the indoor temperature affect the suction pressure?

6. What should be done if there is no refrigerant in the system before startup?

7. What type of service valve ports did the unit you worked on have? (service valves or Schrader valves)

8. What is the typical refrigerant used in central air conditioning systems?

Lab 47 Identifying Controls of a Central Air Conditioning System

Name ______________________________ Date ____________ Grade ________

OBJECTIVES: Upon completion of this exercise, you will be able to identify controls for a central air conditioning system and state their function.

INTRODUCTION: You will remove panels from the condensing unit and the indoor air handler, identify each safety control that concerns the airflow for cooling, and describe its function.

TEXT REFERENCES: Unit 26.

TOOLS AND MATERIALS: A flashlight, straight blade and Phillips screwdrivers, and a VOM.

SAFETY PRECAUTIONS: Turn the power off while working inside the control compartments. Verify with the VOM that the power is off. Lock and tag the panel or disconnect box and keep the key with you.

PROCEDURES

1. What type of air handler are you working with? Gas, oil, electric furnace, or a fan section only ________
2. With the power off and locked out, remove the cover from the fan section and read on the motor the type of motor protection. Record the type here.________ What type of circuit protection is provided for the air handler? fuse or breaker ________ Record the rating of the circuit protection here. ________ A
3. Replace all parts to the air handler with the correct fasteners.
4. With the power off and locked out, remove the compressor compartment panel. Follow the suction line from where it enters the unit to where it enters the compressor and look for a low-pressure control. Reviewing the wiring diagram may help. Is there a low-pressure control? yes or no ________
5. Follow the discharge line from where it leaves the compressor to where it enters the condenser and look for a high-pressure control. It may be located on the liquid line between the condenser and the line leaving the cabinet. Does the unit have a high-pressure control? yes or no ________
6. Examine the line voltage control for an overload for the compressor. Review the wiring diagram. Does the unit have an overload? yes or no ________ Based on the compressor nameplate, does the compressor have internal motor protection? yes or no ________ What type of circuit protector does the unit have? fuse or breaker ________
7. Replace the panels with the correct fasteners.

MAINTENANCE OF WORK STATION AND TOOLS: Return all tools to their places. Leave the work area clean and orderly.

SUMMARY STATEMENT: Describe how a compressor with an internal overload and internal relief valve has low charge protection.

QUESTIONS

1. What is the difference between a fuse and a breaker?

2. Why does a fuse only protect the circuit when applied to an air-cooled condensing unit?

3. How is the low-voltage circuit normally protected?

4. What is impedance protection for a motor?

5. How does an internal relief valve work inside a compressor?

6. How would a low-pressure control located in the liquid line in the condenser protect a system when it may be set to shut off at 10 psig?

7. What is the difference between a fuse and a time delay fuse?

8. What is the difference between a thermal motor protector and a current-type motor protector?

9. How is a typical hermetic motor cooled?

10. What can electronic controls do that electromechanical controls cannot do?

LAB 48 Evaporator Operating Conditions

Name ______________________________ Date ______________ Grade ________

OBJECTIVES: Upon completion of this exercise, you will be able to state typical evaporator pressures and describe the results of reduced airflow across an evaporator.

INTRODUCTION: You will use a gage manifold and thermometer to measure and then record pressures and temperatures under normal conditions in an evaporator in an air conditioning system. You will then reduce the airflow and measure and record pressures and temperatures again.

TEXT REFERENCES: Unit 27.

TOOLS AND MATERIALS: A gage manifold, thermometer, gages, goggles, gloves, cardboard for blocking the evaporator coil, straight blade and Phillips screwdrivers, and 1/4" and 5/16" nut drivers.

SAFETY PRECAUTIONS: Wear gloves and goggles while connecting and removing gages. Be careful not to come in contact with any electrical connections when cabinet panels are off.

PROCEDURES

1. Place a thermometer sensor in the supply and return air streams.
2. With gloves and goggles on, fasten the low- and high-side gages to the gage ports.
3. Start the unit and record the following after the unit and conditions have stabilized.
   - Suction pressure __________ psig, refrigerant boiling temperature __________°F, inlet air temperature __________°F, outlet air temperature __________°F, discharge pressure __________ psig
4. Reduce the airflow across the evaporator to approximately one-fourth by sliding the cardboard into the filter rack. If there is a single return inlet, the cardboard may be placed across three-fourths the opening.
5. When the unit stabilizes, record the following:
   - Suction pressure __________ psig, refrigerant boiling temperature __________°F, inlet air temperature __________°F, outlet air temperature __________°F, discharge pressure __________ psig
6. Remove the air blockage and gages and return the unit to normal operation.

MAINTENANCE OF WORK STATION AND TOOLS: Replace all panels with proper fasteners. Return all tools to their places.

SUMMARY STATEMENT: Describe how the unit suction pressure and air temperature differences responded.

QUESTIONS

1. How does a fixed bore expansion device respond to a reduction in load, such as when the filter is restricted?

2. How does a thermostatic expansion valve respond to the above condition?

3. What are the results of small amounts of liquid reaching a hermetic compressor?

4. How does the temperature drop across an evaporator respond to a reduced airflow?

5. What happens to the head pressure when a reduced airflow is experienced at the evaporator?

6. List 4 things that can happen to reduce the airflow across an evaporator.

7. What procedure may be used to thaw a frozen evaporator?

8. What would happen if an air conditioner were operated for a long period of time without air filters?

9. How may an evaporator coil be cleaned?

LAB 49 Condenser Operating Conditions

Name ______________________________ Date ______________ Grade ________

OBJECTIVES: Upon completion of this exercise, you will be able to describe how a condenser responds to reduced airflow.

INTRODUCTION: You will measure and record a typical condenser's refrigerant pressure and temperature, the condenser entering air stream temperature, and the compressor amperage under normal operating conditions. You will then reduce the airflow across the condenser, measure and record the same conditions, and compare the results.

TEXT REFERENCES: Unit 27.

TOOLS AND MATERIALS: A thermometer, clamp-on ammeter, gage manifold, goggles, gloves, sheet plastic or cardboard to reduce the airflow through the condenser, straight blade and Phillips screwdrivers, and 1/4" and 5/16" nut drivers.

SAFETY PRECAUTIONS: Wear goggles and gloves while attaching and removing gage lines. Be very careful not to come in contact with any electrical terminals or moving fan blades.

PROCEDURES

1. On an air conditioning system that uses R–22 refrigerant, and with goggles and gloves on, fasten the gage lines to the gage ports.
2. With the unit off, fasten one thermometer lead to the discharge line on the compressor and place the other in the air stream entering the condenser. Replace the compressor compartment panel if the compressor compartment is not isolated from the condenser.
3. Clamp the ammeter on the compressor common terminal wire.
4. Start the unit and allow it to run until stabilized (when the temperature of the discharge line does not change). This may take 15 minutes or more. Record the following information:
   - Suction pressure __________ psig, refrigerant boiling temperature __________°F, discharge pressure __________ psig, refrigerant condensing temperature __________°F, compressor current __________ A
5. Cover part of the condenser coil with the plastic or cardboard until the discharge pressure rises to about 300 psig for R–22 and allow the unit to run until it stabilizes, about 15 minutes. Then record the following information:
   - Suction pressure __________ psig, refrigerant boiling temperature __________°F, discharge pressure __________ psig, refrigerant condensing temperature __________°F, compressor current __________A
6. Remove the cover and allow the unit to run for about 5 minutes to reduce the head pressure.
7. Shut the unit off and remove the gages.
8. Replace all panels with the correct fasteners. Return the unit to its normal condition.

MAINTENANCE OF WORK STATION AND TOOLS: Return all tools to their places.

SUMMARY STATEMENT: Describe how and why the pressures and temperatures responded to the reduced airflow across the condenser.

QUESTIONS

1. Does the compressor work more or less with an increased head pressure?

2. What does the discharge line temperature do when an increase in head pressure is experienced?

3. What would a very high discharge line temperature do to the oil circulating in the system?

4. Name 4 things that would cause a high discharge pressure.

5. How does the compressor amperage react to an increase in discharge pressure?

6. Would the discharge pressure be more or less for a high-efficiency unit when compared to a standard-efficiency unit?

7. How is high efficiency accomplished?

8. Is a high-efficiency unit more or less expensive to purchase?

9. Why do many manufacturers offer both standard- and high-efficiency units?

LAB 50 Manufacturer's Charging Procedure

Name ____ Date ________________ Grade __________

OBJECTIVES: Upon completion of this exercise, you will be able to use a manufacturer's charging chart to check for the correct operating charge for a central air conditioning system.

INTRODUCTION: You will use a manufacturer's charging chart for a suggested charge and charging procedures for top efficiency in a central air conditioning system in outside temperatures above 65°F. This unit may have any type of metering device.

TEXT REFERENCES: Unit 28.

TOOLS AND MATERIALS: A thermometer, psychrometer, gage manifold, two – 8" adjustable wrenches, service valve wrench, cylinder of R–22 refrigerant, flat blade and Phillips screwdrivers, 1/4" and 5/16" nut drivers, light gloves, goggles, tape, insulation (to hold temperature sensor and insulate it), and an air conditioning system with manufacturer's charging chart.

SAFETY PRECAUTIONS: Wear gloves and goggles while fastening gage lines and removing them from the gage ports. Be very careful around rotating components.

PROCEDURES

1. Review the manufacturer's procedures for charging. It may call for a superheat check. NOTE: All manufacturers' literature specifies that the indoor airflow be correct for their charging procedures to work. If the readings you get are off more than 10% and cannot be corrected by adjusting the refrigerant charge, suspect the airflow.
2. With goggles and gloves on, fasten the gage lines to the gage ports (Schrader ports or service valve ports).
3. Start the unit and allow it to run for at least 15 minutes.
4. While the unit is stabilizing, prepare for taking and recording the required readings:
   - Suction pressure reading __________
   - Superheat (suction line temperature – boiling temperature) __________°F
   - Discharge pressure __________ psig
   - Condensing temperature (converted from discharge pressure) __________°F
   - Indoor wet-bulb __________°F, dry-bulb __________°F
   - Outdoor dry-bulb __________°F
5. If the unit has the correct charge because it corresponds to the manufacturer's chart, you may shut the unit off and disconnect the gages.
6. If it does not, under your instructor's supervision, adjust the charge to correspond to the chart. Then shut the unit off and disconnect the gages.
7. Replace all panels with the correct fasteners.

MAINTENANCE OF WORK STATION AND TOOLS: Return all tools to their respective places and be sure your work station is in order.

SUMMARY STATEMENT: Describe how the wet-bulb temperature reading affects the load on the evaporator.

QUESTIONS

1. What would a typical superheat reading be at the outlet of an evaporator on a design temperature day for a central air conditioning system?

2. How does the outdoor ambient temperature affect the head pressure?

3. What is the recommended airflow per ton of air conditioning for the unit you worked on?

4. What would the result of too little airflow be on the readings you took?

5. What type of metering device did the unit you worked with have?

6. Did you have to adjust the charge on the unit you worked with?

7. How would too much charge affect a unit with a capillary tube?

8. Would the overcharge in Question 7 be evident in mild weather?

LAB 51 Charging a System in the Field Using Typical Conditions (Fixed-Bore Metering Device)

Name ______________________________ Date ______________ Grade ________

OBJECTIVES: Upon completion of this exercise, you will be able to correctly charge a system using a fixed-bore metering device when no manufacturer's literature is available.

INTRODUCTION: You will check the charge in a central air conditioning system that uses R–22 under field conditions when no manufacturer's literature is available. This system must have a fixed-bore metering device. Always use manufacturer's literature when it is available. The following procedure may be done in any outdoor ambient temperature; however, the colder it is, the more condenser cover you will need.

TEXT REFERENCES: Unit 28.

TOOLS AND MATERIALS: A thermometer, gage manifold, tape, insulation, gloves, goggles, two 8" adjustable wrenches, service valve wrench, plastic or cardboard to restrict airflow to the condenser, flat blade and Phillips screwdrivers, 1/4" and 5/16" nut drivers, cylinder of R–22, and a central air conditioning system with a fixed-bore metering device.

SAFETY PRECAUTIONS: Wear gloves and goggles when attaching and removing gages. Do not let the head pressure rise above 350 psig. Be careful of rotating parts and electrical terminals and connections.

PROCEDURES

1. With goggles and gloves on, fasten the gages to the gage ports.
2. Start the unit and allow it to run for at least 15 minutes for it to stabilize. During this time, fasten the thermometer to the suction line at the condensing unit and insulate it, Figure 1.
3. Observe the pressures and cover the condenser to increase the head pressure to 275 psig, if necessary. Keep track of it and do not let it rise above 275 psig. Regulate the pressure until it stabilizes at exactly 275 psig. NOTE: If it will not rise to 275 psig, the unit may not have enough charge. Check with your instructor. The superheat should be 10°F to 15°F at the condensing unit with a correct charge and an average length suction line.
4. Adjust the charge to the correct level. Do not add refrigerant too fast or you will overcharge the system. It is easier to adjust the charge while adding refrigerant than while removing it.
5. When the charge is correct, record the following:
   - Suction pressure __________ psig
   - Suction line temperature __________°F
   - Refrigerant boiling temperature (converted from suction pressure) __________°F
   - Superheat (suction line temperature – refrigerant boiling temperature) __________°F
   - Recommended superheat for line set length __________°F
   - Discharge pressure __________ psig
   - Condensing temperature __________°F
6. When you have adjusted the charge, remove the head pressure control (cardboard or plastic) and allow the unit to stabilize. Then record the following information:
   - Suction pressure __________ psig
   - Suction line temperature __________°F
   - Refrigerant boiling temperature (converted from suction pressure) __________°F
   - Superheat (suction line temperature – refrigerant temperature converted from suction pressure) __________°F
   - Superheat (suction line temperature – refrigerant boiling temperature) __________°F

7. Shut the unit off, remove the temperature sensors and gages and replace all panels with the correct fasteners.

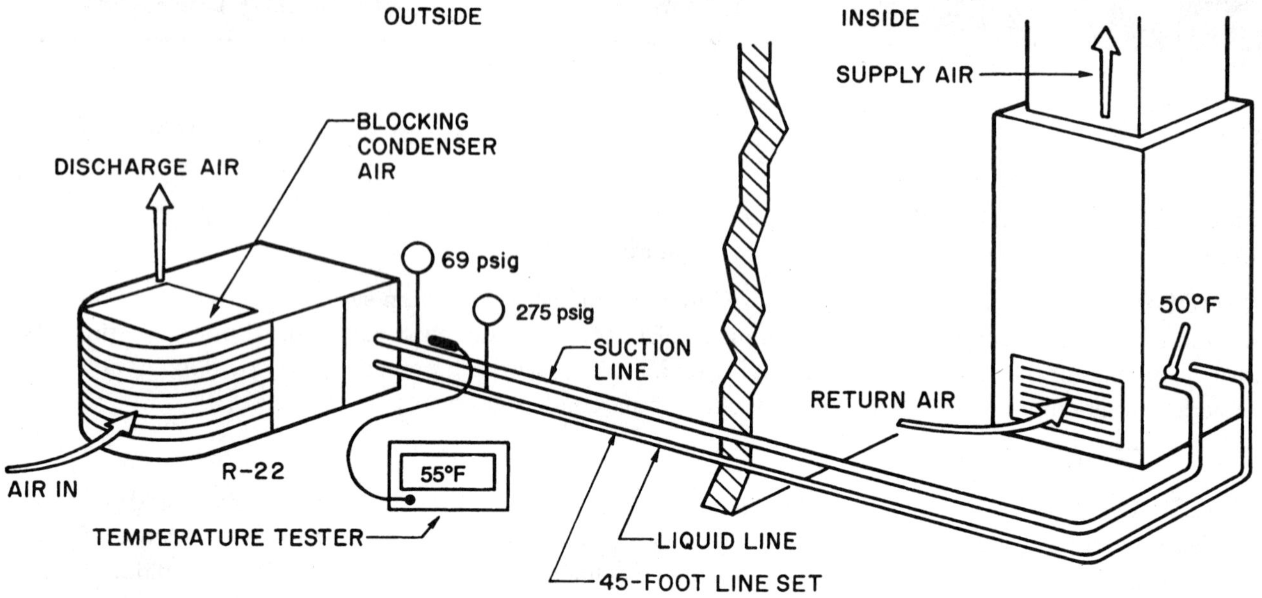

Figure 1

MAINTENANCE OF WORK STATION AND TOOLS. Return all tools, equipment, and materials to their storage places.

SUMMARY STATEMENT: Describe how head pressure affects superheat with a fixed-bore metering device.

QUESTIONS

1. How does line length affect refrigerant charge?

2. Why is the head pressure adjusted to 275 psig instead of some other value?

3. What is the typical refrigerant used for central air conditioning?

LAB 52 Charging a System in the Field Using Typical Conditions (Thermostatic Expansion Valve)

Name ______________________ Date ______________ Grade ________

OBJECTIVES: Upon completion of this exercise, you will be able to correctly charge a system using a thermostatic expansion valve metering device when no manufacturer's information is available.

INTRODUCTION: You will use the subcooling method to assure the correct charge for the unit. This procedure is to be used when there is no manufacturer's literature available and may be performed in any outdoor ambient temperature. A unit with R–22 is suggested because it is the most common refrigerant; however, a unit using other refrigerants may be used.

TEXT REFERENCES: Unit 28.

TOOLS AND MATERIALS: A thermometer, gage manifold, R–22 refrigerant, goggles, light gloves, plastic or cardboard (to restrict airflow through condenser), two 8" adjustable wrenches, a service valve wrench, flat blade and Phillips screwdrivers, 1/4" and 5/16" nut drivers, tape, and insulation for the temperature sensor.

SAFETY PRECAUTIONS: Wear gloves and goggles while fastening and removing gages. Do not let the head pressure rise above 350 psig. Be careful of rotating components and electrical terminals and connections.

PROCEDURES

1. Fasten the gage lines to the test ports (Schrader or service valve ports). Locate the sight glass in the liquid line if there is one.
2. Start the unit and while it is stabilizing, fasten the thermometer sensor to the liquid line at the condensing unit and insulate it, see Figure 1.
3. Place the condenser airflow restricter in place and allow the head pressure to rise to 275 psig. Watch closely, particularly if the weather is warm, to make sure that the head pressure stays at 275 psig. Keep adjusting the plastic or cardboard until 275 psig is steady. If it will not reach 275 psig, look to see if there are bubbles in the sight glass. The unit is probably undercharged.
4. Record the following information:
   - Suction pressure __________ psig
   - Discharge pressure __________ psig
   - Condensing temperature (converted from head pressure) __________ °F
   - Liquid line temperature __________ °F
   - Subcooling (condensing temperature – liquid line temperature) __________ °F
5. If the subcooling is 10°F to 15°F, the unit has the correct charge. If it is more than 15°F, the unit has too much refrigerant. If less than 10°F, the unit charge is too low.
6. Adjust the charge to between 10°F and 15°F and record the following again:
   - Suction pressure __________ psig
   - Discharge pressure __________ psig
   - Condensing temperature (converted from head pressure) __________ °F
   - Liquid line temperature __________ °F
   - Subcooling (condensing temperature – liquid line temperature) __________ °F

7. When the charge is correct, remove the temperature sensor, gages, and replace all panels with the correct fasteners.

MAINTENANCE OF WORK STATION AND TOOLS. Return all tools, equipment, and materials to their storage places and make sure your work station is clean and in order.

SUMMARY STATEMENT: Describe subcooling and where it takes place.

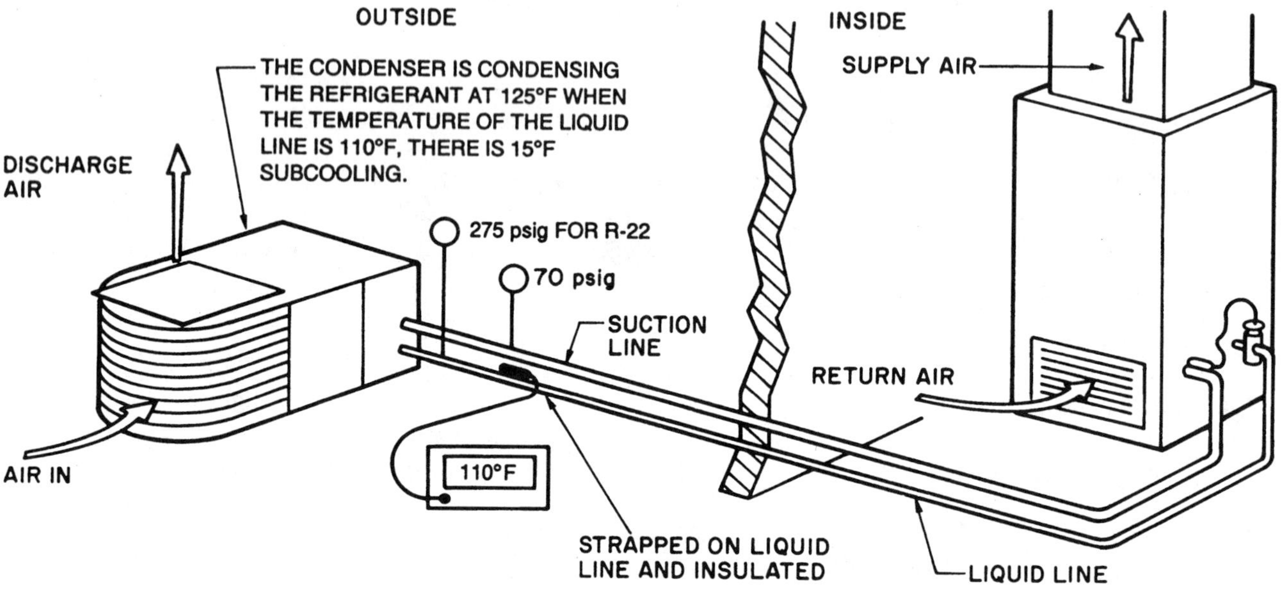

Figure 1

QUESTIONS

1. If the unit you worked on had a sight glass, did it show bubbles when the charge was correct? What caused the bubbles?

2. Where is the correct location for a sight glass in a liquid line in order to check the charge?

3. How is subcooling different from superheat?

4. Does superheat ever occur in the condenser?

5. Why do you raise the head pressure to 275 psig to check the subcooling of the condenser in the above exercise?

6. Why do you insulate the temperature lead when checking refrigerant line temperatures?

LAB 53 Troubleshooting Exercise, Central Air Conditioning with Fixed-Bore Metering Device

Name ______________________________ Date ______________ Grade __________

OBJECTIVES: Upon completion of this exercise, you will be able to troubleshoot a basic mechanical problem in a typical central air conditioning system.

INTRODUCTION: This exercise should have a specific problem introduced to the unit by your instructor. You will locate the problem using a logical approach and instruments. Instruments may be used to arrive at the final diagnosis, however, you should be able to determine the problem without instruments. THE INTENT IS NOT FOR YOU TO MAKE THE REPAIR, BUT TO LOCATE THE PROBLEM AND ASK YOUR INSTRUCTOR FOR DIRECTION BEFORE COMPLETION.

TEXT REFERENCES: Unit 28.

TOOLS AND MATERIALS: Flat blade and Phillips screwdrivers, 1/4" and 5/16" nut drivers, thermometer, two 8" adjustable wrenches, goggles, gloves, and a set of gages.

SAFETY PRECAUTIONS: Use care while connecting gage lines to the liquid line. Wear goggles and gloves.

PROCEDURES

1. Start the system and allow it to run for 15 minutes.
2. Touch test the suction line leaving the evaporator. HOLD THE LINE TIGHT FOR TOUCH TESTING. Record your opinion of the suction line condition here.
3. Touch test the suction line entering the condensing unit and record your impression here.
4. Remove the compressor compartment door. MAKE SURE THAT AIR IS STILL FLOWING OVER THE CONDENSER. IF AIR PULLS IN OVER THE COMPRESSOR WITHOUT A PARTITION, HIGH HEAD PRESSURE WILL OCCUR VERY QUICKLY. COMPLETE THE FOLLOWING VERY QUICKLY AND SHUT THE DOOR. Touch test the compressor body all over. Record the following;
   A. The temperature of the top and sides of the compressor in relation to hand temperature.
   B. The temperature of the compressor crankcase compared to your hand temperature.
5. All of these touch test points should be much colder than hand temperature if the system has the correct problem. Record your opinion of the system difficulty here.
6. Fasten gages to the gage ports.
7. Fasten a thermometer lead to the suction line at the condensing unit.
8. Record the superheat here. __________°F
9. Place a thermometer lead in the inlet and outlet air streams to the evaporator and record their readings here. Inlet air temperature __________°F, Outlet air temperature __________°F, Temperature difference __________°F
10. What should the temperature difference be? __________°F

11. Describe your opinion of the problem and the recommended repair here.

12. Ask your instructor before making the repair, then do as directed.

MAINTENANCE OF WORK STATION AND TOOLS. Leave the unit as directed, return all tools to their respective places.

SUMMARY STATEMENTS: Describe how to determine the problem with this unit by the touch test.

QUESTIONS

1. How should the touch test of the evaporator suction line feel when a unit has an overcharge of refrigerant?

2. How would the compressor housing feel when a unit has an overcharge of refrigerant?

3. How would the suction line feel when a unit has a restricted airflow?

4. How would the compressor housing feel when a unit has a restricted airflow?

5. How can you tell the difference in an overcharge and a restricted airflow by the touch test?

6. Why should gages not be fastened to the system until the touch test is completed?

LAB 54 Troubleshooting Exercise, Central Air Conditioning with Fixed-Bore Metering Device

Name ______________________________ Date ______________ Grade ________

OBJECTIVES: Upon completion of this exercise, you will be able to troubleshoot a mechanical problem in a central air conditioning system with a fixed-bore metering device.

INTRODUCTION: This exercise should have a specific problem introduced to the unit by your instructor. You will locate the problem using a logical approach and instruments. The problem in this exercise will require instruments to properly solve and diagnose. THE INTENT IS NOT FOR YOU TO MAKE THE REPAIR, BUT TO LOCATE THE PROBLEM AND ASK YOUR INSTRUCTOR FOR DIRECTION BEFORE COMPLETION.

TEXT REFERENCES: Unit 28.

TOOLS AND MATERIALS: Flat blade and Phillips screwdrivers, 1/4" and 5/16" nut drivers, gloves, goggles, two 8" adjustable wrenches, thermometer, cylinder of refrigerant, and a gage manifold.

SAFETY PRECAUTIONS: Use care while fastening gage lines to the liquid line. Wear gloves and goggles.

PROCEDURES

1. Start the system and allow it to run for 15 minutes.
2. Touch test the suction line leaving the evaporator. HOLD THE LINE TIGHTLY IN YOUR HAND. Record the temperature of the line in comparison to hand temperature.
3. Touch test the suction line entering the condensing unit. Record the temperature of the line in comparison to hand temperature.
4. Remove the compressor compartment door for the purpose of touch testing the compressor. NOTE: IF THE COMPRESSOR COMPARTMENT IS NOT ISOLATED, THIS TEST MUST BE PERFORMED QUICKLY OR THE HEAD PRESSURE WILL BECOME TOO HIGH DUE TO AIR NOT FLOWING OVER THE CONDENSER. LOOK THE SYSTEM OVER BEFORE LEAVING THE DOOR OFF TOO LONG. Touch test the compressor and record:

   A. The top and sides of the compressor housing compared to hand temperature.
   B. The bottom, crankcase, of the compressor. Replace the door quickly if air is bypassing the condenser.

5. What is your opinion of the problem with this unit using the touch test for diagnosis?
6. Fasten gages to the gage ports.
7. Place a thermometer lead on the suction line entering the condensing unit and record the superheat here __________°F
8. Take the temperature difference across the evaporator by placing a thermometer lead in the inlet and outlet air. Inlet air temperature __________°F, Outlet air temperature __________°F, Temperature difference __________°F
9. What should the temperature difference be? __________°F

10. Record your opinion of the unit problem and your recommended repair.

11. Consult your instructor as to what to do next.

MAINTENANCE OF WORK STATION AND TOOLS. Return the unit to the condition your instructor directs you and return all tools to their respective places.

SUMMARY STATEMENT: Describe how overcharge effects the capacity of a fixed-bore metering device system.

QUESTIONS

1. How does liquid refrigerant affect the oil in the crankcase of a compressor?

2. What is the difference in liquid slugging in a compressor and a small amount of liquid refrigerant entering the compressor crankcase?

3. Do most compressors have oil pumps?

4. How is a compressor lubricated if it does not have an oil pump?

5. Describe how a suction line feels if liquid refrigerant is moving toward the compressor?

6. Do all central air conditioning systems use fixed-bore metering devices?

7. Name two types of fixed-bore metering devices.

LAB 55 Changing a Hermetic Compressor

Name ______________________________ Date ______________ Grade ________

OBJECTIVES: Upon completion of this exercise, you will be able to change a compressor in a typical air conditioning system.

INTRODUCTION: You will recover the refrigerant from a system, then you will completely remove a hermetic compressor from the air conditioning unit and replace it as though you had a new compressor. You will then leak check, evacuate, and charge the unit with the correct charge.

TEXT REFERENCES: Unit 28.

TOOLS AND MATERIALS: Flat blade and Phillips screwdrivers, 1/4" and 5/16" nut drivers, torch set-up suitable for this exercise, solder, flux, slip joint pliers, VOM, ammeter, refrigerant, leak detector, gloves and goggles, a set of socket wrenches, and needlenose pliers.

SAFETY PRECAUTIONS: Use care while working with liquid refrigerant. Turn all power off before starting to remove the compressor. Lock and tag the panel or disconnect box where the power is turned off. There should be a single key; keep it on your person.

PROCEDURES

1. Recover the refrigerant from the system using an approved recovery system.
2. With the refrigerant out of the system, TURN OFF AND LOCKOUT THE POWER.
3. Remove the door to the compressor compartment.
4. Remove the compressor mounting bolts or nuts.
5. Mark all electrical connections before removing. USE CARE REMOVING THE COMPRESSOR TERMINALS AND DO NOT PULL ON THE WIRE. USE THE NEEDLE NOSE PLIERS AND ONLY HOLD THE CONNECTOR.
6. Disconnect the compressor refrigerant lines. Normally you would use the torch to heat the connections and remove them. Your instructor may instruct you to cut the compressor suction and discharge lines leaving some stubbs for later exercises.
7. Remove the compressor from the condenser section completely. You now have the equivalent of a new compressor, ready for replacement.
8. Place the compressor on the mounts.
9. Reconnect the compressor suction and discharge lines in the way your instructor tells you.
10. When the refrigerant lines are reconnected, pressure the system enough to leak check. NOTE: YOUR INSTRUCTOR MAY WANT YOU TO ADD A FILTER DRIER. THIS WOULD BE NORMAL PRACTICE WITH A REAL CHANGE OUT.
11. Evacuate the system. While evacuation is taking place, fasten the motor terminal connectors and the motor mounts.

12. When evacuation is complete, add the correct charge, either by weight or volume.

13. Place the ammeter on the common wire of the compressor and the voltmeter across the line side of the contactor and turn the power on.

14. Ask your instructor to look at the system, then start the compressor and observe the amperage and voltage. Compare with the nameplate.

15. Let the unit run for 15 minutes and check the charge, using the touch test to be sure refrigerant is not flooding over into the compressor.

16. When you are satisfied all is well, turn the unit off and remove the gages.

MAINTENANCE OF WORK STATION AND TOOLS: Replace all panels and return the tools to their places.

SUMMARY STATEMENT: Describe the three steps of triple evacuation.

QUESTIONS

1. Why is it necessary to recover the refrigerant rather than just remove it to the atmosphere?

2. Why should you always check the voltage and amperage after a compressor changeout?

3. Should you always add a filter drier when changing a compressor?

LAB 56 Changing an Evaporator in an Air Conditioning System

Name ______________________ Date ______________ Grade ________

OBJECTIVES: Upon completion of this exercise, you will be able to change an evaporator on a central air conditioning system.

INTRODUCTION: You will recover the refrigerant from a central air conditioning system and change the evaporator coil. You will pressure check and evacuate the system, then add the correct charge. NOTE: IF THE SYSTEM HAS SERVICE VALVES, YOU MAY PUMP THE REFRIGERANT INTO THE CONDENSER.

TEXT REFERENCES: Unit 28.

TOOLS AND MATERIALS: Flat blade and Phillips screwdrivers, 1/4" and 5/16" nut drivers, two 8" adjustable wrenches, set of sockets, gloves, goggles, liquid line drier, cylinder of refrigerant, and gages.

SAFETY PRECAUTIONS: Wear goggles and gloves while transferring liquid refrigerant.

PROCEDURES

1. Put on goggles and gloves, fasten gages to the gage ports, and purge the gage lines.
2. Either pump the refrigerant into the condenser or recover it with an approved recovery system.
3. Turn the electrical power off and lock and tag the disconnect box. With the system open to the atmosphere, sweat or loosen the fitting connecting the suction line to the evaporator. Then loosen the liquid line.
4. Remove the condensate drain line. Plan how this is done so it can be replaced leak free.
5. Remove the evaporator door or cover.
6. Slide the evaporator out.
7. Examine the evaporator for any problems. Look at the tube turns. Clean it if needed. IF YOU CLEAN IT, BE CAREFUL NOT TO ALLOW WATER TO ENTER THE COIL. TAPE THE ENDS.
8. Examine the condensate drain pan for rust and possible leakage. You may want to clean and seal the drain pan while it is out.
9. Examine the coil housing for air leaks. Seal them from the inside if the coil housing is in a positive pressure to be sure it is leak free.
10. When you are sure all is well, slide the coil back into its housing.
11. Reconnect the suction and liquid line.
12. Purge the system with nitrogen, then install the liquid line drier.
13. Pressure the system for leak checking purposes and leak check all connections.
14. Evacuate the system.
15. While evacuation is taking place, install the condensate drain line and replace the coil cover.

16. When evacuation is complete, charge the system, using weight or volume.

17. With your instructor's permission, start the system.

18. Allow the system to run for at least 15 minutes, then check the charge, using the touch test.

19. Turn the system off.

MAINTENANCE OF WORK STATION AND TOOLS: Return the system like the instructor directs you and return all tools to their places.

SUMMARY STATEMENT: Describe why the system was purged before adding the liquid line drier.

QUESTIONS

1. What will a liquid line filter-drier remove from the system?

2. What will the vacuum pump remove from the system?

3. Why is it recommended to measure the charge into the system?

4. Why would an evaporator ever need changing?

5. Is it necessary to remove an evaporator to clean it properly if it is real dirty?

6. Name 3 styles of evaporator.